GEOGRAPHIC INFORMATION SYSTEMS

Geographic Information Systems (GIS) are now essential to the work of an increasingly diverse range of organizations concerned with population characteristics, from local government and health authorities to major retailers and marketing agencies.

The period since the mid-1980s has seen massive growth in the field of GIS – computer-based systems for the handling of geographically referenced information. This growth is due to an explosion in the availability of geographically referenced datasets in the public and commercial domains. Many countries held major censuses of population at the start of the 1990s and the resulting datasets have initiated new waves of GIS applications and specialist software development. GIS have a unique role in the integration of the existing records held by government and commerical organizations, and form the focus for a growing software and data industry.

This book provides a clear and concise overview of GIS development, theory and techniques which addresses the needs of the student or professional who must understand and use GIS for the first time. The reader is taken step by step through the collection, input, storage, manipulation and output of data in GIS, with frequent reference to diagrams and examples. Unique in its focus on the socioeconomic applications, this book allows the reader to develop a strong position from which to question and judge the validity of contemporary GIS technology and literature.

David Martin is Reader in Geography at the University of Southampton.

GEOGRAPHIC INFORMATION SYSTEMS

Second edition

Socioeconomic applications

David Martin

London and New York

First published 1996
by Routledge
11 New Fetter Lane, London EC4P 4EE

Simultaneously published in the USA and Canada
by Routledge
29 West 35th Street, New York, NY 10001

Typeset in Garamond by
J&L Composition Ltd, Filey, North Yorkshire

Transferred to Digital Printing 2003

British Library Cataloguing in Publication Data
A catalogue record for this book is available from the British Library

Library of Congress Cataloguing in Publication Data
A catalogue record for this book has been requested

ISBN 0–415–12571–5
0–415–12572–3 (pbk)

For Caroline and Eleanor: more special still

CONTENTS

PLATES

The following plates appear between pages 90 and 91.

1 Percentage of population under 10 years of age in the City of Southampton (1991 Census)
2 MOSAIC lifestyle groups in West London: a contemporary neighbourhood classification scheme
3 1991 enumeration-district boundaries and 100m postcode locations for Cardiff study area
4 1991 enumeration-district boundaries, street centrelines and ADDRESS-POINT locations for Cardiff study area
5 Population density surface model of Greater London (1991 Census)
6 Population density surface model of Greater London (inset) (1991 Census)
7 Cartogram representation of ward populations with no occupation (1991 Census)
8 Discrete populated settlements in the Greater London region

FIGURES

TABLE

PREFACE TO THE SECOND EDITION

While writing this second edition, it has been gratifying to see that many of the observations and predictions of the first edition have largely held true, but there have also been many unforeseen developments which make a new edition necessary. Simple historical events such as the new population censuses of the early 1990s in most countries have had a major impact on the use of GIS for socioeconomic applications, increasing digital data availability and user awareness, in the context of a maturing software and data industry. GIS have become a familiar part of many undergraduate geography programmes, although there is still uncertainty as to the extent of their centrality in the study of geography. More than ever before, hardware constraints are ceasing to be the limiting factors for most GIS users and national policy and organizational issues are becoming increasingly dominant. The GIS literature has also grown enormously in the four years since the first edition, yet no one else has chosen to adopt the particular approach taken here. There is still much evidence that the early years of GIS development were concerned with modelling and monitoring the physical environment, and that there is a place for a distinct approach to the subject grounded in human geography. One of the most exciting technical trends is progress (admittedly slow) towards greater analytical capabilities within GIS software, and the new possibilities for geographic research to which these may lead. It is also very apparent that GIS use is at every stage subject to the values and conventions of the user: just as with paper maps in past generations, the ownership of GIS and spatial data confers power. We cannot justify the treatment of GIS as only a 'technical' subject, and this realization is also incorporated into the revised discussion.

The overall structure of this edition is largely similar to the first, but a number of changes have been necessary in order to accommodate new information and the emergence of some important new themes. More international examples have been given, and an attempt has been made to make general discussions more relevant to the situations encountered outside the UK. Many of the concepts and principles introduced here are relevant to all GIS users, but the primary focus of this text is still unashamedly on socioeconomic applications, as this was one of the distinctive features of the first edition, which was welcomed by many readers. In particular, there are new sections throughout dealing with major new datasets and developments which have occurred in the early 1990s. Chapter 4 (theories of GIS) has seen the inclusion of some new discussion concerning the role of GIS in the wider field of geographic enquiry, and Chapters 8 and 9 (data output and socioeconomic applications) have seen significant rearrangement and updating, giving particular attention to visualization, and to the value-laden nature of GIS activity. Wherever possible, references to key concepts which had originally appeared in conference proceedings and trade journals have been replaced with more recent references from books and journals, which are more likely to be found in university and college libraries. One certain prediction is that the speed of change will itself continue to increase, and for this reason, detailed examples which refer to specific software systems have again been avoided, as they become outdated so quickly.

David Martin
Southampton, Christmas 1994

ACKNOWLEDGEMENTS

Thanks are due to numerous people who have influenced my ideas since the first edition of this book, and who in various ways encouraged me to write a second. Many of them have also contributed specific information from their own specialist knowledge. In particular I should mention many colleagues at the Department of Geography in Southampton, including Neil Wrigley, Mike Clark, Paul Curran, Ted Milton and Richard Gascoigne. Away from Southampton, my continued work with Paul Longley and Gary Higgs has provided ongoing GIS application experience, proving that it is often easier to write about it than to put it into practice! Finally, my approach to this work owes a great deal to the late Ian Bracken: a good friend, whose knowledge and enthusiasm were always a source of new ideas and encouragement.

The Census data used in the preparation of Plates 1, 5, 6, 7 and 8 and Figure 9.6 are © Crown Copyright. The digital boundaries used in Plates 1, 3 and 4 are Crown and ED-LINE Copyright, and were provided as part of the ESRC/JISC 1991 Census of Population Initiative. The data reproduced in Plate 2 are Copyright © CCN Marketing 1992, © Post Office 1992 and © Automobile Association 1992; the ADDRESS-POINT data in Plate 4 are used by kind permission of Ordnance Survey © Crown Copyright; Plate 7 is reproduced from *Environment and Planning A* with the permission of Pion Ltd and Daniel Dorling. Assistance with illustrations was provided by Paul Cresswell and the Cartographic Unit at the University of Southampton.

GLOSSARY OF COMMON GIS TERMS AND ACRONYMS

ACORN	A classification of residential neighbourhoods. One of the first commercially produced neighbourhood classifications in the UK
ACSM	American Congress on Surveying and Mapping
AGI	Association for Geographic Information: set up in the UK following the Report of the Chorley Committee (DoE, 1987) to coordinate the activities of the GIS industry
AMF	Area Master File: street-network data file (Canada)
ASPRS	American Society for Photogrammetry and Remote Sensing
Attribute data	Any non-spatial characteristic of an object (e.g. total population of a zone, name of a street)
Auto Carto	(i) Automated Cartography (ii) A series of international conferences on all aspects of digital mapping and GIS
Bit	A single binary digit: the smallest unit of digital data
BLPU	Basic Land and Property Unit: contiguous area of land under uniform property-ownership rights
BSI	British Standards Institute
BSU	Basic spatial unit: smallest geographic object, used as building blocks of a georeferencing system
Byte	Unit of digital data, usually that required to store a single character, now almost always eight bits
CAC	Computer-assisted cartography
CAD	Computer-aided design
Cadastral	Concerned with property ownership, particularly for taxation purposes
CAM	Computer-aided mapping (in other contexts, computer-aided manufacturing)
CGIS	Canada Geographic Information System: one of the earliest major GIS installations

Chip	Small fragment of semiconductor (typically silicon) on which an integrated circuit is constructed
Choropleth	Mapping using shaded areas of equal value
CLI	Canada Land Inventory
CPD	Central Postcode Directory: contains 100m grid references for all UK unit postcodes, also known as the Postzon file
CPU	Central processing unit: core component of computer, carries out instructions provided by the software
Dasymetric	Mapping (e.g. population distribution) by use of additional (e.g. land-use) information
Database	A collection of related observations or measurements held within a computer system
Data model	Rationale for a particular data organization scheme within a database
Data structure	The strategy adopted for the organization of data held in a computer system; generally used to refer to more technical aspects than data 'model'
DBMS	Database management system: software to control the storage and retrieval of integrated data holdings (see also RDBMS)
Delaunay triangulation	Triangulation of a set of points such that each point is connected only to its natural neighbours; used in construction of Dirichlet tessellation and as basis for TIN data structures
DEM	Digital elevation model: data structure for representation of a surface variable, usually the land surface (also DTM = digital terrain model)
Digitize	Encode into digital (i.e. computer-readable) form, usually applied to spatial data; hence digitizing tablet
DIME	Dual Independent Map Encoding (US Census data)
Dirichlet tessellation	Complete division of a plane into Thiessen polygons around data points, such that all locations closer to one point than to any other are contained within the polygon constructed around that point
DLG	Digital line graph: a widely used vector data format
DN	Digital number: term applied to the intensity value of a single pixel in a remotely sensed image
DoE	Department of the Environment (UK)
DTM	see DEM
Earth observation	see Remote sensing
ED	Enumeration district: smallest areal unit of UK Census of Population

File	A collection of related information stored in a computer: e.g. a program, word-processed document or part of a database
GAM	Geographical Analysis Machine: an approach to the automated analysis of point datasets without the prior definition of specific hypotheses
GBF	Geographic Base File: street-network-based data file for census mapping (US)
Geocode	Any reference which relates data to a specific location, e.g. postcode, grid reference, zone name.
Geodemography	Techniques for the classification of localities on the basis of their socioeconomic characteristics, typically for a commercial application such as direct mail targeting or store location
Geographic data	Data which define the location of an object in geographic space, e.g. by map coordinates; implies data at conventional geographic scales of measurement
Georeference	Means of linking items of attribute information into some spatial referencing system (e.g. a postcode)
GIS	Geographic information systems
GP	General practitioner: physician providing general health care (UK)
GPS	Global positioning system: series of satellites from which portable ground receivers can calculate their precise locations anywhere on the earth's surface
GUI	Graphical user interface: computer interface based on graphical concepts, i.e. windows, icons, mouse and pointer
HAA	Hospital Activity Analysis: management information system widely used in NHS hospitals
Hardware	Physical computer equipment: screens, keyboards, etc.
HMLR	Her Majesty's Land Registry (UK)
IBIS	Image Based Information System: an early US GIS installation
IFOV	Instantaneous field of view: term used to describe the ground resolution of a remote sensing scanner
Image processing (IP)	Manipulation and interpretation of raster images, usually the product of satelite remote sensing
Internet	Literally 'inter-network': used to refer to the worldwide coverage of interconnecting computer communications networks
I/O	Input/output: computer operation concerned with input or output of information to the central processing unit

Isolines	Lines on a map representing equal values (e.g. contours)
Isopleth	Mapping using lines of equal value (especially of density)
Kb	Kilobyte: i.e. 2^{10} (1,024) bytes
Kernel	In interpolation techniques, a window on which local estimation is based
Kriging	A spatial interpolation technique, also known as 'optimal interpolation', originally developed in the mining industry
LANDSAT	Series of five earth observation satellites, frequently used as data source for IP and GIS
LBS	Local Base Statistics (UK 1991 Census of Population): detailed tabulated statistics for local areas at ward level and above
Line	Spatial object: connecting two points
LINMAP	Line-printer mapping: an early mapping program designed to create cartographic output on a line printer
LIS	Land information system: a GIS for land-resources management
LPI	Land and Property Identifier: addressable object name which refers to a single BLPU (UK)
MAUP	Modifiable areal unit problem
Mb	Megabyte: i.e. 2^{20} (1,048,576) bytes
Memory	That part of a computer concerned with the retention of information for subsequent retrieval
MER	Minimum enclosing rectangle: pairs of coordinates which describe the maximum extents in x and y directions of an object in a spatial database
Metadata	'Data about data': information describing the coverage, origins, accuracy, etc. of datasets
MOSAIC	A commercially available neighbourhood classification scheme (UK)
MSS	Multispectral scanner: electromagnetic scanner carried by earlier LANDSAT satellites
NCGIA	National Center for Geographic Information and Analysis: US research centre for GIS
Network	(i) Physical configuration of cables etc., and associated software, allowing communication between computers at different locations (ii) Data structure for route diagrams, shortest-path analysis, etc.
NHS	National Health Service (UK)
NJUG	National Joint Utilities Group (UK)

Node	Spatial object: the point defining the end of one or more segments
NOMIS	National Online Manpower Information System: an online database of UK employment information, accessible to registered users from remote sites
NORMAP	One of the earliest mapping programs to produce output using a pen plotter
NTF	National Transfer Format
Object-oriented	Originally a programming language methodology, which is based around the definition of types of 'objects' and their properties, increasingly seen as a powerful structure for handling spatial information
OPCS	Office of Population Censuses and Surveys (UK census office)
OS	Ordnance Survey (UK national mapping agency)
OSGR	Ordnance Survey grid reference
OSNI	Ordnance Survey Northern Ireland
OSTF	Ordnance Survey Transfer Format
PAC	Pinpoint Address Code: a commercially produced directory giving 1m grid references for each postal address in the UK
PAF	Postcode Address File: directory of all postal addresses in the UK, giving corresponding unit postcodes
PAS	Patient Administration System in NHS hospitals
PC	Personal computer
PIN	A commercially available neighbourhood classification scheme (UK)
Pixel	Single cell in a raster matrix, originally 'picture element'
Point	Spatial object: discrete location defined by a single (x,y) coordinate pair
Polygon	Spatial object: closed area defined by a series of segments and nodes
Program	An item of software, comprising a sequence of instructions for operation by a computer
Quadtree	Spatial data structure, based on successive subdivision of area, which seeks to minimize data redundancy
Raster	Basis for representation of spatial information in which data are defined and processed as cells in a georeferenced coverage (a raster)
RDBMS	Relational database management system: a strategy for database organization, widely used in GIS software
Remote sensing (RS)	Capture of geographic data by sensors distant from the phenomena being measured, usually by satellites and aerial photography

RRL	Regional Research Laboratory: UK centres for regional research with strong GIS emphasis
SAS	(i) Small Area Statistics (UK Census of Population) (ii) A widely used computer package for statistical analysis
SASPAC	Small Area Statistics Package: widely used software for the management and retrieval of UK Census data for small areas (SASPAC91 for 1991 Census)
SDA	Spatial data analysis
Segment	Spatial object: a series of straight line segments between two nodes
SIF	Standard Interchange Format
SMSA	Standard Metropolitan Statistical Area (US)
Software	Operating systems and programs
Spatial data	Data which describe the position of an object, usually in terms of some coordinate system, may be applied at any spatial scale
SQL	Structured query language: a standard language for the retrieval of information from relational database structures
SYMAP	Synagraphic mapping system: one of the earliest and most widely used computer mapping systems, producing crude line-printer output graphics
Thiessen polygons	Polygons produced as the result of a Dirichlet tessellation
TIGER	Topologically Integrated Geographic Encoding and Referencing System: developed for 1990 US Census of Population
TIN	Triangulated irregular network: a variable-resolution data structure for surface models, based on a Delaunay triangulation
TM	Thematic mapper: electromagnetic scanner used in later Landsat satellites
Unix	Multi-user computer operating system, particularly common on workstations
USBC	United States Bureau of the Census
USGS	United States Geological Survey
UTM	Universal Transverse Mercator: a map projection
VDU	Visual display unit: device for ephemeral visual output from computer (screen)
Vector	Basis for representation of spatial information in which objects are defined and processed in terms of (x,y) coordinates.
VICAR	Video Image Communication and Retrieval: an early US image-processing system

ViSC Visualization in scientific computing: growing use of scientific data exploration using high-power interactive computer graphics

Workstation Configuration of hardware in which a single user has access to a local CPU, screen(s), keyboard, and other specialized equipment; popular environment for GIS operation

ZIP code US five-digit postal code; also ZIP Plus 4: more geographically detailed code with four additional digits

1

INTRODUCTION

This book is intended to provide a general introduction to the field of geographic information systems (GIS). More specifically, it is aimed at those who have a particular interest in the socioeconomic environment. In the following chapters, the reader will find an explanation of GIS technology, its theory and its applications, which makes specific reference to data on populations and their characteristics. The aim is not to provide detailed analysis of fields such as database management systems or computer graphics, which may be found elsewhere, but to introduce GIS within a strong framework of socioeconomic applications. As will be seen, many early GIS applications related primarily to the physical environment, both natural and built. Much writing on the subject reflects these themes, and is unhelpful to the geographer or planner whose interest lies in the growing use of GIS for population-related information. The unique issues raised by these new developments form the specific focus of this text. This introduction sets the scene, both in terms of GIS and of the socioeconomic environment. Some concepts appearing here with which the reader may be unfamiliar will be addressed in more detail as they are encountered in the text.

GEOGRAPHIC INFORMATION SYSTEMS

Geographic information, in its simplest form, is information which relates to specific locations. Figure 1.1 illustrates four different types of geographic information relating to a typical urban scene. The physical environment is represented by information about vegetation and buildings. In addition, there are aspects of the socioeconomic environment, such as bus service provision and unemployment, which cannot be observed directly, but which are also truly 'geographic' in nature. The 1980s saw a massive rise in interest in the handling of this information by computer, leading to the rapid evolution of systems which have become known as 'GIS'. It must be stressed, however, that the use of digital data to represent geographic patterns is not new, and only in the last decade has 'GIS' become a

1

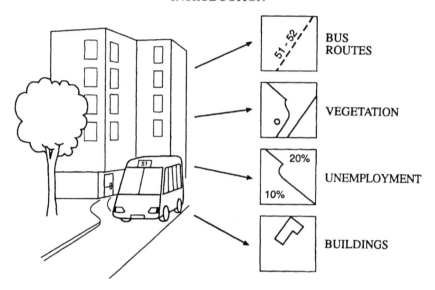

Figure 1.1 Examples of geographic information

commonly used term. Nevertheless, confusion exists as to what exactly constitutes a GIS, and what functions it should perform. While the commercial world is able to offer systems with ever-increasing functionality, there are still no generally accepted theoretical guidelines for their application. Certainly, there has been no clear theoretical structure guiding the developments which have taken place, largely in response to specific user needs. This lack of theoretical work has been a common criticism in much writing about the field (Berry, 1987; Goodchild, 1992).

The systems which we now call 'GIS' have grown out of a number of other technologies and a variety of application fields, and are thus a meeting point between many different disciplines concerned with the geographic location of their objects of study. GIS potentially offer far greater power for manipulation and analysis of data than had been available with earlier systems, broadly aimed at map or image reproduction, but also place greater demands on data accuracy and availability. The data in GIS may be accessed to obtain answers to questions such as 'what is at location X ?', 'what areas are adjacent to route R ?' or 'how many of object A fall within area B ?', which is true geographic 'information'. A major theme which will run through these discussions is the way in which GIS provide an accessible and realistic model of what exists in the real world, allowing these kinds of questions to be addressed. This is very much more than merely using a computer to 'draw maps', and is potentially a very powerful tool, but it also provokes questions as to the nature of the 'real' world, and forces us to consider whose definitions of 'reality' we are going to adopt. The

2

implications of these developments for geography are far-reaching, as they offer a technology which can dynamically model some aspects of the geographic world. The need to examine carefully the mechanisms by which this is achieved is therefore fundamental. A geographic information system, then, according to these criteria, may be summarized as having the following characteristics:

1 *Geographic*. The system is concerned with data relating to geographic scales of measurement, and which are referenced by some coordinate system to locations on the surface of the earth. Other types of information system may contain details about location, but here spatial objects and their locations are the very building blocks of the system.

2 *Information*. It is possible to use the system to ask questions of the geographic database, obtaining information about the geographic world. This represents the extraction of specific and meaningful information from a diverse collection of data, and is only possible because of the way in which the data are organized into a 'model' of the real world.

3 *System*. This is the environment which allows data to be managed and questions to be posed. In the most general sense, a GIS need not be automated (a non-automated example would be a traditional map library), but should be an integrated set of procedures for the input, storage, manipulation and output of geographic information. Such a system is most readily achieved by automated means, and our concern here will be specifically with automated systems.

As suggested by this last point, the data in a GIS are subject to a series of transformations and may often be extracted or manipulated in a very different form to that in which they were collected and entered. This idea of a GIS as a tool for transforming spatial data is consistent with a traditional view of cartography, and will be used here to help structure the discussion of the concepts of GIS.

A brief review of the diverse academic and commercial literature would suggest that typical application areas have been land resources and utility management, but applications in the fields of census mapping and socio-economic modelling have experienced massive growth following the new censuses of the early 1990s, and there has been a corresponding increase in large in-house databases compiled by many organizations. The absence of any firm theoretical basis means that we have no mechanisms for evaluating the appropriateness of these diverse applications. This is of particular importance in the context of growing interest in the field, as the existing theoretical work is broadly descriptive, and cannot offer much help in any specific application. Evaluation of GIS installations has been largely in terms of the financial costs and benefits of replacing traditional procedures. The breadth of potential applications has made GIS a subject of government and research council interest in the UK and USA and there is currently much demand for training in the use of these systems. In 1987, the Committee of Enquiry into the Handling of Geographic Information in Britain published its influential report, commonly referred to as the 'Chorley Report' (Department of the Environment, 1987). This report touched on many of the issues which will be raised here, particularly illustrating the importance of spatially referenced socioeconomic information.

In drawing together submissions for its investigations, the Chorley Committee also demonstrated the difficulties faced by those wishing to obtain clear information about GIS, because of the diverse applications and piecemeal development mentioned above. Much of the relevant literature existed in the form of conference proceedings and papers in technical journals relating to specific applications. It was not until 1987 that an international academic journal concerned purely with GIS emerged. The last five years have seen an enormous explosion in the GIS literature with additional GIS-related academic journals, a number of international trade journals and many new textbooks, but the commercial nature of the industry has led to the continuance of a heavy reliance on a non-academic literature with limited availability. Available textbooks now range from large collections of papers covering all aspects of GIS (such as Maguire *et al.*, 1991) to those concerned with very specific application areas (such as Haines-Young and Green, 1993). This diversity of literature still poses difficulties for those concerned with socioeconomic information who may have little familiarity with the earlier application fields.

References to the operation of specific software systems have been avoided here as far as possible, due to the speed with which such information dates. It is a feature of the buoyancy of GIS development that software suppliers are constantly developing new modules to handle particular aspects of specialized data processing, meet new international transfer and processing standards, or take advantage of some new range of workstation (for example), that the situation changes with each new software release. The topics chosen for discussion are felt to be sufficiently well established and enduring that they will aid the reader in understanding the important aspects of specific systems with which they may come into contact.

SOCIOECONOMIC DATA

An important and fast-developing application field for GIS is that of socioeconomic or population-related data. By population-related data we are here referring to data originally relating to individual members of a population which is scattered across geographic space, such as the results of censuses and surveys and the records gathered about individuals by health authorities, local government and the service sector. These data are here contrasted with those which relate to physical objects with definite locations, such as forests, geological structures or road networks. Clearly, there are relationships between these types of phenomena, as they both exist in the same geographic space, but in terms of the data models with which we shall be concerned, it is important to recognize the distinction between them.

A number of census mapping systems with broadly GIS-type functionality and data structures have evolved. GIS are now being used extensively for health care monitoring, direct mail targeting and population mapping (Hirschfield *et al.*, 1993; Birkin, 1995) However, data relating to dynamic human populations are very different in their geographic properties to those relating to the physical world: the location of any individual is almost always referenced via some other spatial object, such as a household address or a census data-collection unit. Unlike a road intersection or a mountain summit, we are rarely able to define the location of an individual simply by giving their map reference. This has far-reaching implications: socioeconomic phenomena such as ill health, affluence and political opinion undoubtedly vary between different localities, but we cannot precisely define the locations of the individuals which make up the chronically sick, the affluent or the politically militant. If GIS are to be used to store and manipulate such data, it is crucial that much care is given to ensuring that the data models used are an acceptable reflection of the real world phenomena. Different interest groups may have very different conceptions of which phenomena are important and what is 'acceptable'. Again, the absence of a theoretical structure makes it difficult to identify the nature of the problems, and the levels at which they need to be addressed.

The importance of these issues in many countries is increased by a growing interest in geodemographic techniques which has been fuelled significantly as the results of the censuses of 1990 (e.g. USA, France) and 1991 (e.g. UK, Portugal) have become available, together with much more data about the individual collected by a wide range of organizations (Beaumont, 1991; Arnaud, 1993; Thrall and Elshaw Thrall, 1993). There continues to be growth in the demand for computer systems to manipulate and analyse all these new sources of information, and GIS geared specifically towards the handling of socioeconomic data are multiplying rapidly. In the light of these developments and the technology- and application-led growth of GIS to date, it is essential that a clear theoretical framework be established and that the particular issues relating to socioeconomic data types be understood. These are the issues which this book seeks to address, and we will often return to consider their implications during this introduction to GIS.

STRUCTURE OF THE BOOK

Chapters 2 and 3 are essentially historical, tracing the development of GIS through two closely allied technologies, computer-assisted cartography and image processing, and going on to consider the present role of GIS. Some introductory remarks on computers and information systems should provide the essentials for those with no computing experience, but it is assumed that the reader will look elsewhere for more detailed explanation of computers and their operation. Chapter 2 demonstrates the largely ad hoc development of GIS, and introduces the concept of the information system as a way of modelling certain aspects of the real world. This idea is central to the picture of GIS technology portrayed in later chapters. Chapter 3 places these technological developments in context by giving attention to some typical GIS applications and demonstrating the environments in which these systems have been used and developed. Particular attention is given to application examples involving socioeconomic data, and the chapter illustrates the powerful influences of computer capabilities, and physical-world applications on the evolution of contemporary systems.

Chapter 4 reviews the existing theoretical work on GIS, identifying two main themes. The first of these, the 'components of GIS' approach, is based around the functional modules which make up GIS software. The second theme, termed the 'fundamental operations of GIS' seeks to view GIS in terms of the classes of data manipulation that they are able to perform. Neither approach offers much help to those wishing to transfer GIS technology to new types of data, and an alternative framework is presented which focuses on the characteristics of the spatial phenomena to be represented and the transformations that the data undergo. These are identified as data collection, input, manipulation and output. The transformational view of data processing is consistent with current theoretical work in computer cartography (Clarke, 1990), and develops the differences between GIS and merely cartographic operations. This discussion serves to identify further the unique characteristics of socioeconomic phenomena over space. A grasp of these conceptual aspects is seen as essential for a good understanding of the ways in which GIS may be used, and the correct approach to developing new applications. A theme which runs throughout this text is the importance to socioeconomic GIS applications of having a clear conceptual view of what is to be achieved. This is in clear contrast to much of the existing work. Chapter 4 also addresses the extent to which GIS are contributing to the development of a separate 'geographic information science'.

The four important transformation stages identified in Chapter 4 form the basis for the discussion which follows. Chapters 5 to 8 explain how GIS provide a special digital model of the world, looking at each of the data transformations in turn, and giving particular attention to data storage. In

Chapter 5, data-collection and input operations are considered. This includes an overview of some of the most important sources of spatially referenced socioeconomic information, with particular reference to the way in which these data are collected and the implications for their representation within GIS. Knowledge of the processes of data collection is very important to ensure the valid manipulation of the data within GIS, and this issue is addressed. The discussion then moves on to examine the techniques for data entry into GIS, looking at vector, raster and attribute input methods. This section includes data verification, stressing the difficulties involved in obtaining an accurate encoding of the source data in digital form.

The various structures used for the organization of data within GIS form a very important aspect of their digital model of the world, and different data structures have important implications for the types of analyses which can be performed. In addition to the common division into vector and raster strategies, object-oriented approaches and triangulated irregular networks are considered. There are a variety of approaches to the encoding and storage of information, even within these different approaches, and these are addressed in Chapter 6, which also explains some of the most important aspects of attribute database management.

The feature of GIS which is most commonly identified as separating them from other types of information system is their ability to perform explicitly geographic manipulation and query of a database. In terms of the data transformation model, these operations define the difference between GIS and cartographic systems, and they are examined in Chapter 7. Examples of the application of GIS technology to socioeconomic data manipulation are illustrated, including neighbourhood classification and matching census and postcode-referenced databases. It is the provision of such powerful tools as data interpolation, conversion and modelling which makes GIS so widely applicable and of such importance to geographic analysis.

The final aspect of GIS operation is the output of data in some form, either to the user or to other computer information systems. Chapter 8 explains the principles of data display and transfer, addressing both the technology used and the need for commonly accepted standards. Developments in data availability and exchange will continue to be particularly influential on the path of GIS evolution, both by attracting new users and by limiting the interchange of information between organizations. In both aspects, the available technology is considerably more advanced than the user community, and the need for good graphic design principles and more open data exchange is explained. This discussion also considers 'the power of GIS' in terms of the ways in which GIS tend to use selected information about the world in order to reinforce the position of certain groups in society. Debate is only just beginning to emerge in this important area concerning the appropriate use of GIS technology and its role within academic investigation.

Following the introduction to the development and techniques of GIS, there is a need to reconsider how this technology may be applied to the representation of the socioeconomic environment. Chapter 9 focuses specifically on the use of GIS mechanisms to develop systems for modelling the socioeconomic environment. Different approaches are evaluated, which are based on individual-level, area-aggregate and modelled surface concepts of the population and its characteristics.

Each chapter begins with an overview, which seeks to set the chapter in its context in the more general framework of the book, and to preview the topics which will be considered. Chapters conclude with a short summary section which reinforces the main principles relevant to the discussion elsewhere. An additional deterrent faced by the novice reader of GIS literature is the apparently impenetrable mass of acronyms and buzz-words. A glossary of many of the terms and acronyms used here is provided at the beginning of the book, although each of these is also explained in the text where it first appears. The reader is also referred to more general GIS glossaries such as that provided by Green and Rix (1994).

2

THE DEVELOPMENT OF GIS

OVERVIEW

In this chapter, we shall consider the ways in which GIS technology has developed, tracing the relevant influences in other information systems that are also concerned with the representation of geographic data. It is important to realize that the evolution of GIS has taken place over a long period of time which has itself been a period of rapid evolution and growth in all aspects of computing. It is beyond the scope of this book to give a history of these developments, or even to provide an adequate introduction to this enormous field, but the next two sections do at least seek to show how GIS fit into the overall picture of information technology and will provide some basic knowledge for the complete beginner.

The development of GIS cannot be viewed in isolation from two other important areas of geographic information handling by computer, namely computer-assisted cartography (CAC), and remote sensing and image processing. These are both distinct technologies in their own right, with large numbers of commercially available computer systems, established literatures, and histories at least as long as that of GIS. Each has made significant contributions to the field which we now call GIS, and is therefore deserving of our attention here. By examining briefly the areas covered by each of these related technologies, we shall increase our understanding of what comprises a GIS and the kinds of problem to which it may be applied. The emphasis given to CAC and image processing here reflects our primary concern with the representation of spatial data, rather than the technical development of the systems used. A distinction which will become apparent between these contributory fields and GIS is the way in which the data are organized in GIS to provide a flexible model of the geographic world. The convergence of these two fields and their implications for GIS were acknowledged in the mid-1980s by the House of Lords Select Committee on Science and Technology (Rhind, 1986).

It must be stated at the outset that there is no clearly agreed definition of when a computer system is or is not a GIS: but it is possible to identify the key characteristics which distinguish these systems from others, and it is around this view that the rest of the book will be structured. This 'identity crisis' of GIS is likely to persist until there is much wider use of such systems and the role of geographic information in organizations becomes more established. To demonstrate this point, Maguire (1991) contains a diverse collection of definitions of GIS selected from the recent literature. Two major themes may be seen running through the recent development of GIS. These are (1) an explosion in the quantities of geographically referenced data collected and available to a wide range of organizations, and (2) increasingly rapid advances in the technologies used for the processing of these data. Geographic information is important to a vast number of decision-making activities, and large quantities of data are often necessary to meet the needs of any particular institution. The complex systems which have developed to process the data required in these situations generally reflect the characteristics of the data which they were designed to handle. This is an idea to which we shall return in some detail later on.

After a review of the contemporary situation regarding GIS development, attention is given in the following chapter to the types of application field in which these technologies have been used. It will become clear from these introductory remarks, the extent to which the available systems have until very recently been influenced by hardware capabilities and applications in the physical environment. The aim of this text is to focus on the problems and potential for socioeconomic applications, and this theme will recur throughout our review of system development.

COMPUTER SYSTEMS

Before considering GIS in any detail, it is necessary to have at least a basic idea of the way in which computers work. Those who have any familiarity with computers and their basic terminology may safely pass on to the next section on p. 13. However, those with no computer experience will find the following remarks helpful. The novice is strongly urged to seek a fuller introduction to the development and role of computers in geography and planning such as that provided by, for example, Maguire (1989) or Bracken and Webster (1989a).

Computers are programmable electronic machines for the input, storage, manipulation and output of data. The two most basic components of a computer system are the hardware and software. The hardware comprises the physical machinery, such as keyboards, screens, plotters and processing units, whereas the software comprises the programs and data on which the hardware operates.

The structure of computer hardware is basically the same, regardless of the size of the system, as illustrated in Figure 2.1, which shows a typical desktop personal computer (PC). This involves one or more devices for the

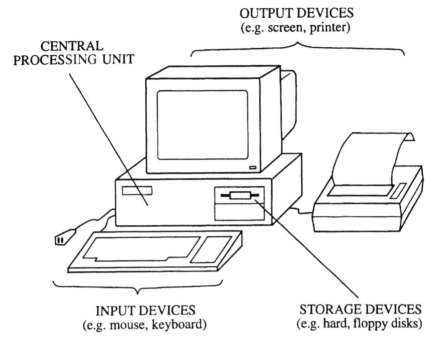

OUTPUT DEVICES
(e.g. screen, printer)

CENTRAL
PROCESSING UNIT

INPUT DEVICES
(e.g. mouse, keyboard)

STORAGE DEVICES
(e.g. hard, floppy disks)

Figure 2.1 Basic hardware components

user to input data and instructions (e.g. a keyboard), and for the machine to output data and responses to instructions (e.g. a screen). In between these two types of device there will be a central processing unit (CPU). This is the 'heart' of the computer and actually represents and processes the data and instructions received from the input devices by means of short electrical pulses on very compact integrated circuits ('chips'). The workings of the CPU and the representation of data in this way need not concern us here, except to add that for most computers, all the information in the machine's electrical 'memory' is lost when the power is switched off. For this reason, in addition to the CPU, we would usually expect to find a device for the storage of these data and instructions when they are not actually in use, and which will preserve them when the machine is switched off. Storage devices usually involve some form of magnetic media on tape or disk using techniques similar to those used by audio or video tape recorders.

Actual systems vary enormously in terms of the number of items of equipment involved and the size and power of their CPUs. Since the introduction of the first electronic computers in the 1950s, there have been constant and accelerating trends towards faster processors and storage devices which are able to handle more data and instructions, and which are available at lower cost and are physically smaller. This progress has been paralleled by massive advances in the quality and sophistication of input/output equipment such as plotters and graphics screens. Increasingly, data on one machine are made available for manipulation by other CPUs by linking them together across networks, similar to those used by telephones and fax machines.

It is now necessary to briefly consider the software component of computer systems. So far, we have carefully avoided reference to programs, but have spoken in terms of the data and instructions which are processed by the computer. As mentioned above, these are assembled and processed as short electrical pulses, which involves breaking down all the familiar letters and numbers of real language into the codes understood by the computer (binary digits, or 'bits'). Computer programs are series of instructions to the machine in special formalized languages (programming languages), which can easily be broken down into the computer's own internal representation codes. This is actually achieved by means of further, specialized programs. Programs of some kind govern all aspects of the machine's operation, a useful distinction being that between the operating system, which governs the running of the machine, and the application programs, which concern the processing of the user's data in some way. Thus an application program may instruct the computer to read data from an input device such as the keyboard, to perform some mathematical operation on those data, and to pass on the result to an output device such as the screen. The encoded data are organized into a system of files and directories, which may be read, edited, copied between disks, accessed by programs, etc. The computer, although possessing no intelligence of its own, is able to perform many tasks based on simple repetitive and logical operations very rapidly, and the art of the computer programmer is to convert the time-consuming and complex problems of the real world into efficient sets of instructions on which the power of the computer may be set to work. Increasingly, the user's interaction with the computer and its software is via a graphical user interface (GUI), in which the programs, data and commands are represented by graphic images on the screen, and interaction is primarily by means of a mouse and pointer, rather than by typing at the keyboard. Although initially associated particularly with Apple's Macintosh range of computers, GUIs of one form or another may now be found across the whole range of machines from the smallest laptop to some of the largest and most powerful computers available. These developments are important to GIS because our interaction with geographic information has always been strongly graphic in nature, involving pointing at maps, tracing routes and the recognition of patterns. Graphic interaction with computers thus has much to offer the GIS user.

INFORMATION SYSTEMS

Computer information systems are essentially no more than sophisticated programs or series of programs designed to represent and manage very large volumes of data about some aspect of the world. Early in the development of computer systems, it was realized that here was a way of automating many operations which had traditionally been performed manually, involving repetitive and error-prone work by specialized human workers.

Examples not involving specifically geographic information include type-setting for traditional printing processes and stock management in a supermarket. In each of these fields, specialized forms of computer information system have been developed, allowing the reproduction of the traditional product very rapidly, and generally with a far greater degree of flexibility than was possible with the traditional manual process. The impact of this technology may be seen in each of the cases cited: word processing and desktop publishing systems allow a user to enter all the information (both text and pictures) to be printed, and to edit and rearrange the page at will, seeing an electronic representation of the finished page on the computer screen, until the desired layout is achieved. The selected information may then be sent directly to the printer, which in this case is another example of an output device which may be attached to the computer. In the second example, that of stock-keeping in a store, there is again an example of the use of a specialized piece of peripheral equipment for the computer, in this case a barcode reader, which identifies the product being sold, and is able to consult a database to discover its price, while at the same time recording its sale, and placing an order for a replacement item to be obtained from the warehouse.

Each of these examples involves the representation of some physical product – in these examples the page of the newspaper or the stock on the shelves – by means of numbers and letters encoded within the computer. In the same way, information systems have developed specially for the representation of items which exist in geographic space. These systems include computer-assisted cartography, in which the many complex operations of conventional cartography are performed within the computer, again using a coded data representation to rapidly reproduce the conventional product in a more flexible way. Image-processing systems also deal with these kinds of data, although encoded rather differently, as we shall see on pp. 21–25. GIS belong to this group of information systems dealing with models of geographic space, and enable the user to answer geographic questions such as 'where is . . . ?' or 'what is at . . . ?' by reference to the data held in the computer rather than having to go out and perform expensive and time-consuming measurements in the real-world. This is, of course, a highly simplified account of what actually happens, but the

13

ability to hold a representation of some real-world system as data in the computer enable us to perform additional operations, which could not be done in reality. A typical example would be the modelling of different future scenarios in order to assess their consequences, simply by manipulating the model in the machine. The potential of the computer information system to replicate or in many cases exceed the capacity of the human typesetter or warehouse manager has tended to reduce the roles of system operators to those of supplying and maintaining the data required by the system, while the previously skilled or repetitive processes are handled by the machine. Clearly, this has major organizational implications for any agency which decides to implement an information system in some aspect of its operation, and a GIS is no exception.

One final distinction is necessary before we consider the development of systems for handling geographic data. It is important to realize that the data models we shall be referring to throughout the rest of this book are for the most part independent of the encoding of data as electrical pulses within the machine, mentioned above. By 'data models' we are referring here to the way in which measurements obtained about the real world are conceptualized and structured within the information system: whether, for example, the boundaries of a parcel of land are considered as a single entity, a series of lines, or a network of lines and points. These issues form the heart of GIS representation of the physical world, and to a large extent determine the usefulness of information systems for answering our questions about the geographic 'real world'. They also prompt certain questions as to the appropriateness of the model's use in any particular context, and the quality and selection of the data supplied to the machine. It is these questions, in the context of the representation of socioeconomic data, which are a major theme of this book.

COMPUTER-ASSISTED CARTOGRAPHY

Computer-assisted cartography (CAC) is an umbrella term used to cover a variety of specialized systems for map and plan creation using a computer. The first of these is the area of computer-aided design (CAD), which is widely used in architectural and engineering design environments and which, when applied to the production of maps, is known as computer-aided mapping (CAM). Within this broad field lie the related aspects of automated cartography, a term generally used for the production of conventional topographic maps by automated means, and the more specialized research activity of statistical and thematic map creation. Due to this confusing array of terms and definitions, Rhind (1977) suggested the general term 'computer-assisted cartography' (CAC), which covers all aspects of map making using computer assistance, and which we shall use here. These systems have had much to do with the development of

14

vector GIS, in which precise spatial coordinate storage and high-resolution graphics have played an important role. The concepts of vector and raster (grid-based) modes of representation, mentioned here, will be discussed fully in Chapter 6.

Monmonier (1982) regards automation as an important step in the evolution of cartography, which he portrays as a line of development stretching back to the earliest recorded maps, and which includes the invention of printing in the fifteenth century. He notes the way in which 'softcopy' map images may replace paper maps in many applications, and outlines potential developments which point towards the integrated geographic information system. This concept has taken many years to realize, although optimistic previews to the new technology have tended to be proved true in the longer term. Moellering (1980) classified the new computer representations of maps as 'virtual maps' which offered far more flexibility than the traditional paper map, which he termed 'real maps'. The new virtual maps had only an ephemeral existence on the computer screen and would avoid the need for large numbers of paper copies. Indeed, the early years of computer cartography were marked by a (perhaps premature) optimism as to the ways in which automation would revolutionize cartography, followed by a period of disappointment and caution on the part of adopting organizations. The early systems were very much oriented towards data display with manipulation capabilities coming only more recently (Green *et al.*, 1985). The extent to which automation has lived up to the initial expectations becomes apparent as we begin to trace its progress.

Rhind (1977) notes that few geographers or professional cartographers were involved in the earliest attempts at drawing maps with computers. Initial developments tended to come from applications in geology, geophysics and the environmental sciences. Although there were suggestions for the use of computers for such cartographic tasks as map sheet layout, name placement and the reading of tabular data such as population registers, the early developments generally represented very low quality cartography. They were designed to produce output on standard line printers, using text characters to produce variations in shading density, as illustrated in Figure 2.2. The most famous of these early programs was SYMAP (SYnagraphic MAPping system), developed at the Harvard Computer Laboratory. SYMAP and LINMAP (LINe printer MAPping), a program written to display the results of the 1966 UK Census of Population, may be seen as representing the first generation of widely used computer mapping systems, and already the influence of hardware capabilities is apparent. SYMAP could produce maps for areas with constant values or interpolate surfaces between data points to produce shaded contour maps for spatially continuous variables (Shepard, 1984), but the line-printer orientation meant maps were limited in resolution, and individual data cells could not be

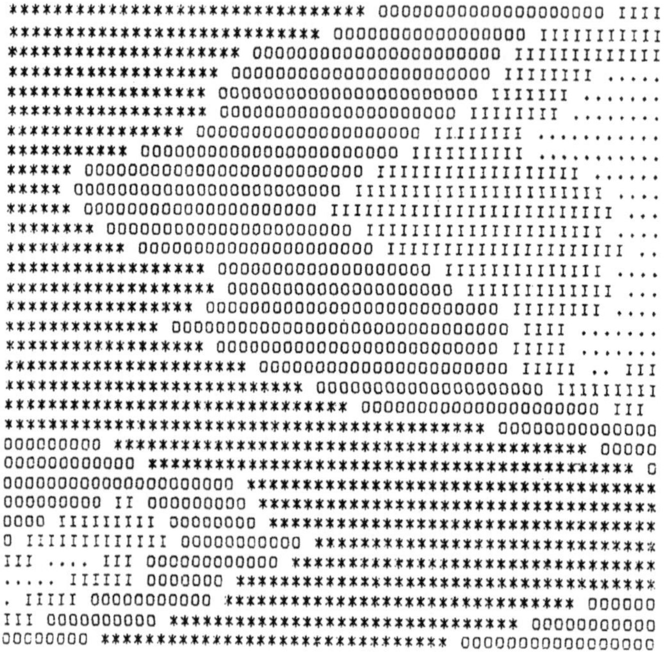

```
****************************** 0000000000000000000 IIII
****************************** 00000000000000000 IIIIIIIIII
******************** 0000000000000000000000 IIIIIIIIIIIII
****************** 00000000000000000000000 IIIIIIII .....
***************** 0000000000000000000000 IIIIII .......
***************** 0000000000000000000000 IIIIIII ........
**************** 00000000000000000000 IIIIIII ...........
********** 000000000000000000000000 IIIIIIII ...........
****** 00000000000000000000000000 IIIIIIIIIIIIIIIII ......
***** 00000000000000000000000000 IIIIIIIIIIIIIIIIIIIIII ....
****** 0000000000000000000000000 IIIIIIIIIIIIIIIIIIIIIIII ...
******* 0000000000000000000000000 IIIIIIIIIIIIIIIIIIIIII ....
********** 00000000000000000000000 IIIIIIIIIIIIIIIIIIIIII ..
***************** 0000000000000000000 IIIIIIIIIIIIIII ....
***************** 00000000000000000000 IIIIIIIIIIIII ...
***************** 0000000000000000000000000 IIIIIIII ....
************** 00000000000000000000000000000 IIII .......
***************** 00000000000000000000000 IIIII .......
********************** 0000000000000000000000 IIIII .. III
************************** 0000000000000000000 IIIIIIII
***************************** 00000000000000000000 III
*********************************************** 0000000000000
000000000 **************************************** 00000
000000000000 ****************************************** 0
0000000000000000000 ****************************************
000000000 II 00000000 ********************************
0000 IIIIIIIII 00000000 **************************
0 IIIIIIIIIII 000000000 **************************
III .... III 000000000 **************************
...... IIIIII 0000000 ***************************
. IIIII 0000000000 **************************** 000000
III 000000000 ************************** 00000000000
0000000 ************************** 00000000000000000
```

Figure 2.2 An example of line-printer map output

represented by square symbols. An advance was the use of coordinate-based information, and the use of pen plotters, such as in the NORMAP package developed in Sweden. This was able to produce maps on either plotter or printer, and much care was given to the methods of map construction, in this case with data based mainly on points. Nordbeck and Rystedt (1972) stress: 'The technique used for the presentation of the finished map forms an important part of this process, but even more important are the methods used for processing the original information and for the actual construction of the map'.

The use of coordinate data and plotter output is the starting point for the use of automated cartography by national mapping agencies such as the Ordnance Survey (OS) in the UK, and the United States Geological Survey (USGS). As the capacity of hardware to store and retrieve digital map data began to increase, the advantages of storing data in this form became more apparent. A geographic database which was revised continually would be available on demand for the production of new editions without recompilation. Generalization and feature-coding algorithms would allow for the creation of hardcopy maps at a variety of scales, and the complex issues of projection and symbology might be separated from the actual task of data storage: the 'real map' (see above) would cease to be the main data store and become merely a way of presenting selected information for the customer's requirements from the comprehensive cartographic database held in the computer. It would even be possible to preview the appearance and design of a new product on a graphics screen before committing it to paper, and production itself could be made much faster once the data were entered into the computer.

Morrison (1980) outlined a three-stage model of the adoption of automation, suggesting that technological change has been an almost constant feature of cartography, but that automation was truly revolutionary. The three stages he identified were as follows:

1 the early 1960s: rapid technical development, but a reluctance to use the new methods and fear of the unknown new technology;
2 late 1960s and 1970s: acceptance of CAC, replication of existing cartography with computer assistance;
3 1980 onwards: new cartographic products, expanded potentials, full implementation.

Various reasons contributed to the slow take-up of CAC. The key aspects included the high initial cost of entering the field, and the need to make heavy investments in hardware and software in an environment where both were evolving rapidly. Some initial installations were disappointing and discouraged other map-producing agencies from taking the first step. Once the decision had been taken to adopt a computer-assisted system, with its staffing and organizational implications, it proved very difficult for agencies to go back to manual working practices. Some of the primary determinants of map quality and form were the spatial references used to locate the geographic objects, the volume and organization of the data, and the quality of output devices available. One of the major obstacles to automation for large-scale producers like the OS was (and still is) the task of transforming enormous quantities of existing paper maps into digital form. The latest developments include surveying instruments which record measurements digitally for direct input to digital mapping systems. These include both conventional ground surveying stations which record all measurements internally for subsequent download to a host computer, and the recent availability of global positioning systems (GPS) (Wells *et al.*, 1986; Gilbert, 1994). GPS receivers, of which many are small enough to be hand-held, obtain fixes from a global network of GPS satellites, from which it is possible to compute their location on the surface of the earth. Centimetre precision is achievable, and these systems also have many important non-surveying applications in navigation and defence. Nevertheless, all previously surveyed data are held in paper map form. Despite various attempts at the automation of digitizing (the transformation of geographic data into computer-readable form), even the most advanced scanning and line-following devices (see p. 86) still require a considerable amount of operator intervention and are not really suitable for many applications. Thus the primary form of data input remains manual digitizing, which is a very labour-intensive and error-prone process. Over a period of two and a half decades of experimental digital mapping (Bell, 1978; Fraser, 1984) it has become apparent that the derivation of all OS map scales from a single digital database is not a practical possibility, due to the very great differences in symbology and degree of generalization between the scales. In the 1980s national mapping agencies were under pressure from a wide range of organizations that wished to use survey data in digital form, and the report of the Chorley Committee (DoE, 1987) again stressed the need for OS to rapidly complete a national digital database. One obstacle remains the implementation of efficient mass digitizing procedures, which, although conceptually simple, proved very difficult to achieve (Rhind et al., 1983). Despite these difficulties, the UK has now virtually completed national digital coverage at smaller mapping scales, and is moving rapidly to a situation in which the main products of the national mapping agency are a series of digital databases (Rhind, 1993). Some map products are available as scanned versions of the conventional paper products, without any geographic data structuring, but all datasets are increasingly seen as commodities. The same issues must be faced by any organizations wishing to automate their existing cartographic operations, whether in the context of CAC or of GIS. A clear picture of the potential 'horrors of automation' is given by Robbins and Thake (1988).

Contemporary CAD and CAM systems provide the capacity to store a number of thematic overlays of information (e.g. roads, water features, contours, etc.), each of which may be associated with a particular symbology. This basic idea is illustrated in Figure 2.3, which shows a set of thematic layers superimposed to produce a map. Some specialized CAD software contains sophisticated editing and three dimensional-modelling functions, while modern CAM systems are more concerned with the ability to vary projection and symbolization of output maps. However, neither have strongly developed attribute data handling capabilities, and generally store spatial information as series of feature-coded points and lines without any associated topological information or area-building ability (Dueker, 1985, 1987). By 'attribute data', we are referring here to the non-locational characteristics of a particular geographic object, such as the population of a zone or the name of a street. This organizational structure means that the data are generally only suitable for use within the type of system to which they were entered, and the geo-relational information required for GIS-type queries (above) is not available. For example, road names would typically be held in a separate text layer (if textual information is handled at all), which can be overlaid on, but has no database link with, the road-centreline layer. Consequently, realignment of a road section would require changes to both centreline and text layers, and no facilities exist for cross-layer manipulation

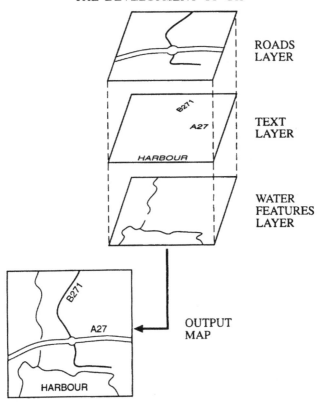

ROADS
LAYER

TEXT
LAYER

WATER
FEATURES
LAYER

OUTPUT
MAP

Figure 2.3 Data layers typical of a CAM system

or analysis, such as the extraction of all centrelines with a particular name. More recently, there have been commercial initiatives towards the integration of some of the leading GIS and CAD software.

It should be noted that socioeconomic data have not figured largely in any of these developments, as the boundaries of census and administrative districts are frequently defined in terms of features belonging to a number of other coverages in a CAC system, and are not in themselves major components of a topographic database. In addition, the inability of such systems to assemble polygon topology or to store multiple attributes for a map feature discourage the use of such systems for census mapping. Another aspect of CAC is automated thematic mapping packages such as MAP91 which are more specifically designed for the production of statistical maps (in this case, census data), and are able to handle area data with multiple attribute fields. This type of map is shown in Figure 2.4, and contrasts with the CAC example in Figure 2.3 (in which each feature is represented only by a symbolized line). The area-based map is made

19

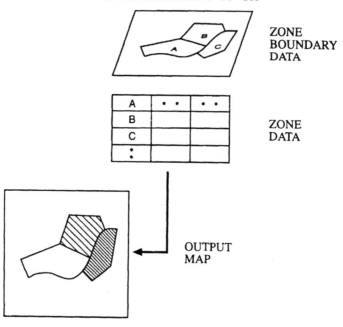

ZONE
BOUNDARY
DATA

ZONE
DATA

OUTPUT
MAP

Figure 2.4 Thematic mapping of zone-based data

possible by the use of more sophisticated data structures, and facilities for the assembly of an area topology from labelled segment information. The different data-structure approaches are examined in detail in Chapter 6. This type of system is usually used for the production of 'one-off' statistical maps and in research situations where there is a need to move between mapped and tabular data. Again, manipulation and spatial analysis functions are limited, and complex statistical manipulation would normally take place in an external system.

Although some automated digitizing and high-quality plotting procedures involve the use of high-resolution raster data, modern CAC is essentially a vector-oriented technology, in which data are held within systems by series of coordinates. This applies particularly in national mapping agencies such as the OS whose primary concern is with the larger map scales. Such information is increasingly available as scanned raster images, but these are suitable only as background information, and are effectively the paper documents stored on computer. Vector databases may sometimes be converted to high-resolution raster information as part of the printing process. Real-world surfaces, such as the elevation of the land surface above sea level, are represented by series of coordinates defining contour lines, which exist as linear features in the database.

IMAGE PROCESSING

The second technology of importance to GIS development has been that of remote sensing (RS) and the activity of image processing (IP) of remotely sensed data. The use of data from satellite and airborne sensors for the monitoring and management of the environment is also increasingly referred to as 'earth observation'. Satellite RS is the largest single source of digital spatial data, and has therefore been a significant factor in the development of mechanisms for fast and efficient processing of these data. The orientation here has been towards raster (image-based) systems, and the entire field has only been made possible by the development of automated systems, which are able to cope with the volumes of data involved. In raster systems, the map is in the form of a grid, each cell of which represents a small rectangular region on the ground, and contains a value relating to that region. Again, these ideas will be explained fully in Chapter 6. Unlike computer-assisted cartography, which represents the automation of a long-established discipline with a theory and conventions of its own, image processing is very much a new technology, a child of the digital spatial data era. In order to understand the rationale behind image processing developments, we must first understand something of satellite remote sensing which is its major data source.

Unlike cartography, RS represents an entirely new field, which may perhaps be traced back to the earliest aerial photography. Massive growth has only been experienced in the last thirty years with the arrival of satellite sensing systems which are returning enormous quantities of continuously sensed digital data. After what Estes (1982) describes as a twenty-year 'experimental stage' in this technology, there is now a far greater focus on the application of RS data, and one of the main avenues for development has been as an input to GIS (Kent et al., 1993). This potential has been acknowledged from the very beginning of RS development, but, as will be seen, technical obstacles have continued to hinder its full realization (Marble and Peuquet, 1983; Michalak, 1993). The attractions of RS data include its potential for uniformity, compatibility within and between datasets, and its timeliness. This may be a means for the maintenance of up-to-date data in GIS systems and an ideal method of change detection by continual monitoring of environmental resources. The following description is highly selective, but more general introductions may be found in Drury (1990) and Barrett and Curtis (1992).

There is a considerable variety of satellites, targeted at different types of earth and atmospheric observation, and employing different sensing devices. Those of particular interest here are the earth resources satellites, generally with sensor ground resolutions of under 0.25km. Satellites designed for earth-resources observation generally have slow repeat cycles, circling the earth perhaps once every two weeks. Clearly, this is too great an interval for weather forecasting, and meteorological satellites have much faster repeat cycles. Satellites may be equipped with photographic or non-photographic sensing devices. Of the earth resources satellites, the longest-running in this context have been the LANDSAT series, with ground resolutions (referred to in the RS literature as 'instantaneous field of view' – IFOV) of 30–80m and more recently the French SPOT satellites, with a resolution down to 10m. Both LANDSAT and SPOT actually carry multiple scanners, which capture electromagnetic information in a variety of wavebands, in the visible and near-visible ranges. This radiation represents the reflectance of the earth of (originally solar) radiation, and the spectral characteristics (the distribution of the radiation across the electromagnetic spectrum) of the reflected radiation at any point is a function of the characteristics of the surface at that point. Thus interpretation of the spectral characteristics of an image should allow us to derive useful information about the surface of the earth. This process is shown diagrammatically in Figure 2.5. LANDSATs 1–3 carried a multispectral scanner (MSS) which recorded data in four wavebands at around 80m resolution, and LANDSATS 4 and 5 carried a more sophisticated scanner known as the thematic mapper (TM), with seven wavebands at up to 30m resolution. LANDSAT 6 was lost soon after launch in 1993, but LAND-SATs 4 and 5 are still in an operational condition, and a seventh satellite in the series is projected for the late 1990s. With MSS and TM one image is produced per waveband per scene, so the output from Landsat 5 for a given area would actually consist of seven images, although not all are likely to be used in a single interpretation exercise. The SPOT platforms have carried a sensor known as HRV (high-resolution visible) with ground resolution of 10–20m, and SPOT-4 is due for launch in 1997. An increasing number of other national and international space programmes have placed earth observation satellites in recent years, with Japanese, Indian, Chinese and European satellites in operation (CEOS, 1992; Curran, 1993). The most recent sensor devices feature very high spectral resolution together with slightly reduced spatial resolution, but there are many potential applications for the extraction of socioeconomic information, particularly in those parts of the globe where routine ground data collection is patchy.

IP systems are the computer systems which have been developed to handle these data. There is an increasing range of 'integrated' IP/GIS systems such as ERDAS and IDRISI which contain a range of functions for image interpretation and manipulation, together with more general-purpose raster GIS abilities. The 'raw' data received from the satellite sensors require considerable radiometric and geometric correction before any attempt can be made at image interpretation, and a number of strategies exist for the interpretation operation itself. Image data consist of a matrix of digital numbers (DN) for each band, each cell (pixel) of

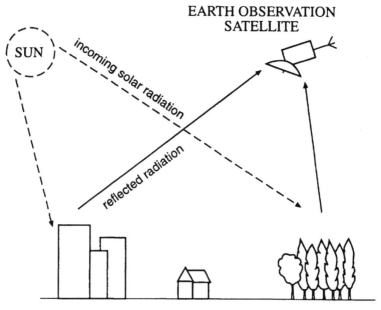

LAND SURFACE (varying reflectance characteristics)

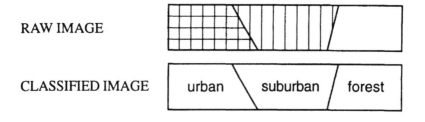

Figure 2.5 Capture and interpretation of a remotely sensed image

which represents the value recorded for the corresponding area on the ground. There is a need to remove known distortions resulting from the curvature of the earth and to suppress the effects of atmospheric scatter and satellite wobble. The values recorded may be affected by skylight and haze and, over large areas, differential degrees of shade and sunlight. Radiometric restoration and correction consists of the adjustment of the output of each detector to a linear response to radiance. Geometric correction involves changing the location of lines of pixels in the image and complete resampling, with reference to ground control points of known locations, in order to georeference the image data. Images may then be enhanced by stretching the distribution of values in each band to utilize the whole 0–255 DN (8-bit) density range, usually in a linear fashion. This is because, in any one image, a relatively small proportion of the sensor measurement scale is used, and image enhancement makes interpretation easier. This enhancement may be performed in a semi-automated fashion, or may be done manually where it is desired to highlight some particular ground features whose spectral characteristics are known. Other enhancement techniques useful in specific situations include a variety of spatial filters which may be used to enhance or smooth images and may be concentric or directional in operation.

Once the image data have been 'cleaned' in this way, the major task remaining is one of image classification. If RS data are to be integrated with existing GIS or digital map data (Trotter, 1991), it is necessary to interpret the DN values and to combine and classify the image bands according to some recognizable classification scheme. This process is illustrated by Figure 2.6, which shows a simplified case using only two bands. 'All remotely sensed images are but a poor representation of the real world' (Curran, 1985), and an intelligent image classification will attempt to equate the DN values with identifiable phenomena on the earth's surface, such as lakes, forests, urban areas, etc. The classification method may be 'unsupervised', in which the IP system performs the entire classification according to a set of predetermined rules without further operator intervention, or 'supervised', in which the operator directs the system in the choice of

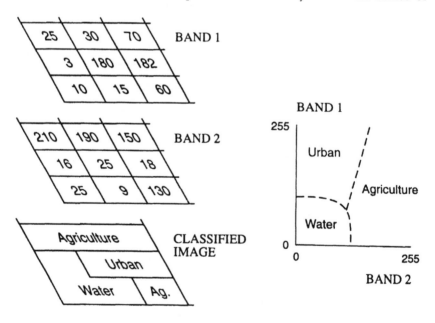

Figure 2.6 Image classification

classes for classification. In either case, classification is conventionally performed on a pixel-by-pixel basis, with reference to the DN for each band in the current pixel. In an unsupervised classification, the system examines all cells and classifies into a given number of classes according to groupings of the image region with similar DN. This technique is known as 'density slicing'. Although reproducible by any operator, there is no guarantee that the classes chosen will bear any relationship to the 'real-world' classes of land cover which we are hoping to discover: it is merely a statistical division of the DN values (Merchant, 1982). Often, there will be an obvious relationship between such a classification and actual land features, as the whole concept of image classification is based on the assumption that different features have different spectral characteristics. However, the classes chosen by an unsupervised classification may not be compatible with any other data existing for the region.

A more common approach is to use a supervised classification in the light of the operator's knowledge of the study region. In such a classification exercise, the operator selects 'training areas' on the image, which are known to be of a particular land-cover type, for as many cover types as are required in the classification. The IP system then groups cells according to the training area whose 'spectral signature' they most closely resemble. Different land-use types may be identified which have consistent and recognizable spectral signatures. Three simple image classes are given in the graph on the right of Figure 2.6. On the basis of this information, all pixels with values of less than 100 in both bands will be classed as water features. Assignment to these classes may be according to minimum distance to means, parallelepiped or maximum likelihood classifiers (Harris, 1987). One of the main problems with either type of classification procedure is that the IP system contains no 'general knowledge' as to what is a sensible classification, and it may be very difficult to distinguish between the spectral characteristics of two very different land uses. More recently, some authors have suggested the use of ancillary or contextual information derived from digital mapping or GIS as an aid to classification (e.g. Mason et al., 1988; Niemann, 1993). Again, socioeconomic interests have not figured prominently in these attempts to record and model geographic reality. Various attempts have been made to use IP for the extraction of settlement patterns and population data from RS imagery, but they have often suffered from a lack of suitable ancillary data to aid classification (Lo, 1989; Sadler et al. 1991).

CONTEMPORARY GIS

Goodchild (1991a) identifies the increasing power of workstations, the spread of networking standards and decreases in hardware costs as important features of the technological setting of GIS. These changes have taken place at all levels from desktop personal computer (PC) to mainframe and made possible many operations within a PC GIS which could never have been contemplated even fifteen years ago. The improvements in networking capabilities and new developments in database architecture are bringing about a situation in which distributed workstations with powerful graphics capabilities work on a mixture of local and remote databases (Webster, 1988; N. P. Green, 1987). This movement has been facilitated by cheaper, higher-resolution graphics, including the availability of sophisticated graphics systems for desktop PCs. The influence of these advances in computing technology on what is practically possible in GIS cannot be overstated, and it must be remembered that the thirty-year existence of GIS and the other spatial data processing systems has been one of enormous advance for computer science itself. Developments have often been uncoordinated, with software designers seeking to utilize the capacity offered by some new device, rather than building from a theoretical base which defines what is really required (Collins *et al.*, 1983). Despite the massive importance of each of these hardware and software fields, they should here be seen as the constantly changing background against which our main theme of spatial representation may be examined.

It will become increasingly apparent the extent to which GIS development has been technology- rather than theory-led, and this is reflected in the nature of the literature. For many years, primary sources of papers have been the proceedings of the conferences of organizations such as Auto Carto (Automated Cartography); the American Society for Photogrammetry and Remote Sensing (ASPRS) and the American Congress on Surveying and Mapping (ACSM). The focus of this work has been very much oriented towards specific technical advances and project applications. In the mid-1990s, the importance of hardware cost and capabilities is being more and more replaced by organizational factors in determining the direction and possibilities for GIS. New international organizations and conference series have emerged for which GIS is the primary focus, such as the European Conferences on Geographic Information Systems (EGIS), yearly since 1990. This period has also seen the emergence of a GIS 'community'. In 1987 the *International Journal of Geographic Information Systems* was founded, the first academic journal concerned solely with GIS developments (Coppock and Anderson, 1987), and in the early 1990s a number of existing academic journals have made shifts of emphasis (and sometimes of title) to explicitly incorporate aspects of GIS. Commercially oriented journals such as *Mapping Awareness, GIS World* and *GIS Europe* have seen considerable growth, filling an important role in communicating research and developments to GIS users, and effectively filling the role of trade journals for the commercial GIS industry. Another measure of the impressive growth of GIS literature is Marshall's (1994) list of no fewer than fifty-five books on GIS of which fifty have been (or were due to be) published in 1990 or later.

The development of the three primary spatial data processing technologies (CAC, IP and GIS) has led to a situation in which distinct commercial markets exist for each type of system, but with increasing integration between the three. A number of cheap PC GIS packages are available, but the more sophisticated of these have tended to be raster-based systems developed in-house by academic and planning agencies and subsequently more widely distributed (e.g. Sandhu and Amundson, 1987; Eastman, 1992). Although the increasing power of personal computers has made sophisticated functionality available in such software, the workstation (see below) has largely continued to be the preferred platform of the major GIS suppliers.

In the development of both software and 'turnkey' systems (in which the purchaser is supplied with hardware, installed software, training and continuing support), recent growth has been particularly focused on the use of networked workstations running under variants of the Unix operating system. Numerous hardware manufacturers are involved in this field, and the leading software may generally be implemented under any of the major suppliers' systems. Common examples of machine series at this level include Sun SPARCstations, Silicon Graphics Indys, IBM RS6000, and Digital DECstations. GIS, with their heavy use of interactive graphics and database access, do not sit comfortably on mainframe machines running many other tasks simultaneously, yet they have database requirements which until very recently have been too large for most PCs: hence the networked workstation environment is well suited to today's applications. We may now be seeing a new trend emerging in which for some installations the most appropriate hardware configuration is a network of the most powerful PC systems, providing 'desktop' GIS for large numbers of users (Maguire and Dangermond, 1994; Rix, 1994). Undoubtedly, the world GIS market leader over recent years has been ESRI's Arc/Info (ESRI, 1993a), but many other suppliers offer broadly comparable systems, including Genasys II, Intergraph and Laserscan. These large systems provide both raster and vector processing modules, relational database architectures, extensive data interfaces to other systems, and the potential for software customization by experienced users. This technology has 'filtered down' from the domain of national-level agencies, which were involved in development of the earliest large systems. The immediate future would appear to be a mixture of desktop and workstation systems, with routine use taking place towards the lower end of the power spectrum and more demanding applications using Unix architecture.

As processing power has become more widely available and GIS more widely known, we are seeing successively smaller organizations making use of GIS. An implication of this trend is that applications are becoming

27

increasingly specialized, although the initial software developed for environmental monitoring still influences the form of today's systems. A substantial literature has developed comprising reviews of specific systems and general experiences of GIS adoption, for the guidance of potential users (American Farmland Trust, 1985; Rhind and Green, 1988; Smith and Tomlinson, 1992). Major themes in earlier reports were the need for trained personnel (Congalton, 1986) and the dangers of misapplication of an expensive technology with significant organizational implications. Increasingly, it is these organizational issues and not technical ones which are determining the success or otherwise of GIS implementation. With the advent of a range of masters-level degree courses in GIS and the inclusion of GIS elements in many university geography programmes, the availability of trained staff has become less problematic. Organizational issues in this context are more likely to include access to and cost of data and the position of GIS within overall IT and information strategies (Dickinson and Calkins, 1988; Smith and Tomlinson, 1992). A common conclusion from this work is that a full understanding of the application issues to which the system is to be addressed is necessary before an appropriate system may be designed. Every application environment has its own unique characteristics which may make one off-the-shelf system more appropriate than others or may require special customization of existing software. A corporate approach to GIS with careful consideration of user needs, data sharing and strong team leadership, seems to be critical to successful implementation.

The influence of public policy on developments in digital mapping has been evident from the earliest implementations (Monmonier, 1983). In the UK, a committee of enquiry was set up under Lord Chorley in 1985 to advise the government 'on the future handling of geographic information . . . taking account of modern developments in information technology' (DoE, 1987: v). The committee published its report in 1987 and the government's official response appeared in 1988 (DoE, 1988). The report stressed the benefits to be obtained from digital data integration, which would involve more work on data standards, and the need for education in the use and application of GIS (Chorley, 1988; Rhind, 1988). A significant feature of the Chorley Report was the extent to which it addressed the handling of socioeconomic data, in addition to that relating to the physical environment, which has been the traditional domain of GIS. This emphasis perhaps reflects a significant difference between the uses for GIS technology in the USA and the UK. In a summary of North American experience, Tomlinson (1987) lists GIS applications by sector, and this classification fails to address socioeconomic data at all. By contrast, a major research initiative in the UK in the late 1980s was the Regional Research Laboratory (RRL) initiative set up by the Economic and Social Research Council (Burrough and Boddington, 1992), reflecting the importance attached to the handling of census and 'geodemographic' data. A number of the RRLs identified the integration of socioeconomic data into GIS technology as a key area of research, and, although the main phase of this initiative is now finished, many of the RRL sites continue to be active in this field.

In the USA, a National Center for Geographic Information and Analysis (NCGIA) was established in August 1988, to form a focus for research in GIS. Many of the research focuses of this centre concern technical issues such as GIS theory, spatial analysis, spatial statistics and database structures, which are of relevance to all application areas, and this is reflected in the published outputs from the initiative (NCGIA, 1989, 1992). Within some of these headings in the centre's research plan, socioeconomic issues are specifically identified, again signalling the increasing awareness of their importance. Also mentioned are issues of data privacy and confidentiality, which are of especial importance where personal data are concerned. Abler (1987) notes a number of differences between the UK and US situations, highlighting in particular the US tendency to allow applications to run well ahead of policy and research. Outside the USA, organizations have tended to be slower to enter the GIS field, and government is more active in promoting and regulating activities, in addition to its much more significant role as a provider of spatial data, especially in the socioeconomic realm. Research agendas at a European level are reported by Arnaud et al. (1993), who describe the GISDATA scientific programme of the European Science Foundation (ESF). Again, there is an emphasis on fundamental research and concern with issues such as data standardization, quality and integration within the European Union.

Another recommendation of the Chorley Report in the UK was that there should be some central body, a 'centre for geographic information', to provide a focus for the many diverse interest groups involved in the handling of geographic information. In its response, the government endorsed the work of existing organizations such as the RRLs, but did not outline any specific proposals for the creation of such a national centre. Subsequent to this, various organizations involved in the field have set up the Association for Geographic Information (AGI) in an attempt to meet this need. The AGI has been active in the organization of major conferences and as a lobbying group representing the interests of geographic data and software users and suppliers. Various international organizations have also emerged at European and global scales, such as the European GIS foundation (EGIS) (Harts et al., 1990).

SUMMARY

In this chapter, we have briefly reviewed the key components of computer systems and introduced the concept of an information system. Information systems are complex software products designed to represent a particular aspect of the 'real' world within the computer, often with the object of more efficiently managing that reality. These systems provide the facility to manipulate and analyse aspects of the data in the model and frequently utilize specialized hardware devices, tailored to particular applications.

Geographic information systems, or GIS, are a special type of information system, concerned with the representation and manipulation of a model of geographic reality. They are closely related to computer-assisted mapping and satellite image-processing systems, which have made significant contributions to GIS technology. These two related technologies have been more specialized in their applications and have tended to lack much of the general analytical power of GIS, although integrated systems are increasingly bringing about a convergence. It is already apparent that those aspects of the 'real' world which are considered appropriate for inclusion in any system are to a large extent system-dependent and will be influenced by the uses for which the system has been established. As our discussion continues, we shall clearly see that GIS do not in any sense provide a definitive objective model of geographic 'reality'.

Enormous advances have been made in general computer hardware and software capabilities, which have facilitated far more sophisticated geographic databases and manipulation possibilities, and without these developments the applied fields would never have become widely used. We are now in a position where organizational issues such as corporate strategy and data access are likely to be more significant than hardware capabilities when considering the feasibility of a potential GIS installation. A review of GIS development in this way reflects the extent to which contemporary systems have for many years been influenced by hardware capabilities and interest in applications to the physical environment. In the following chapter these application fields will be considered in more depth, and the implications of the rapidly growing GIS environment for socioeconomic applications examined more fully.

3

GIS APPLICATIONS

OVERVIEW

In the previous chapter we saw how contemporary GIS have evolved alongside developments in computer-assisted cartography (CAC) and image processing (IP). These technologies have been mostly concerned with monitoring and modelling the physical environment. This was reflected in the orientation of many GIS texts towards land-resources assessment, from Burrough (1986) through to Haines-Young and Green (1993) and Goodchild *et al.* (1993b). This emphasis is also evident in many conference proceedings and collections of papers, where the balance of applications listed falls clearly in the physical domain. It would be impossible to adequately convey all these applications here, although a general distinction has been made between systems primarily concerned with the natural environment and those concerned with the built environment.

In many situations, an organization will have an interest in data which relate both to physical and human aspects of the geographic world. Indeed, given the present increasing awareness of the importance of environmental issues, this need for data integration may be expected to increase. However, the requirements of GIS for handling socioeconomic information are in many cases different from those concerned with phenomena in the physical environment. The emphasis on physical applications in the development of GIS technology has led to a situation in which many presently available systems are not well suited to the modelling of socioeconomic phenomena. Our concern here is primarily with the extension of GIS into the socioeconomic realm, and for this reason disproportionate emphasis is given to existing applications to population and related data. Attention is given to the DIME and TIGER systems, used for the representation of US census data, and to the various systems developed in the UK for handling census-type information. It will become apparent that these applications rely heavily on the conventions established for physical world data. In later chapters, alternative approaches to the representation of such data will be examined. Important influences in the application of GIS technology in the socioeconomic realm are the growing field of geodemography and the increased availability of large population-related datasets. This chapter concludes with an introduction to some geodemographic techniques.

NATURAL ENVIRONMENT

This application field represents some of the largest and longest established GIS installations and probably also represents the most common use of GIS technology. The two historic examples cited here illustrate the use of vector and raster systems, making use of primarily cartographic and remotely sensed data respectively.

The first example is the Canada Geographic Information System (CGIS), begun in 1964, initially to handle information gathered by the Canada Land Inventory (CLI) (Tomlinson *et al.*, 1976). This was one of the first major GIS sites and illustrates many features well ahead of its time. Source data were collected and mapped in polygon (i.e. vector) form, and input either by scanner or manual digitizing (see Chapter 5). Data organization was by thematic 'coverages', such as agriculture, forestry, recreation, land use, census and administrative boundaries, watersheds and shorelines. Descriptive information was held for each zone, with database links to the coded image data. The system actually comprised a collection of programs and subroutines for data retrieval and analysis. A major use of the system was the overlay of zones belonging to different coverages, and the output of information about the new areas defined, primarily in tabular form. Data were input to the system from a variety of sources, mostly as hardcopy map documents previously held by separate agencies, covering the whole of Canada. The system is of special interest because of the sophistication it achieved at a very early stage in the growth of GIS.

Another of the early large systems concerned with environmental monitoring was the Image Based Information System (IBIS) developed at the California Institute of Technology in the mid-1970s (Bryant and Zobrist, 1982; Marble and Peuquet, 1983). The initial use for IBIS was the processing of images obtained from the LANDSAT satellites, and the integration of land use data from the RS images with administrative-area boundaries, captured with a raster scanner. IBIS may be seen as a raster-based GIS, which grew from the Video Image Communication and Retrieval (VICAR) IP system. IBIS was able to take in both tabular and graphical data, including encoded aerial photographs. Integration with the Intergraph CAC system (Logan and Bryant, 1987) increased the flexibility of the system for polygon processing and display, although there are considerable problems with the conversion of data between the vector and raster subsystems. An application in Portland, Oregon (Marble and Peuquet, 1983) incorporated data from LANDSAT, census returns and field measurement of air pollution. The potential for incorporation of population

data is suggested, but problems are encountered in the encoding of census tract boundaries in a way which makes them compatible with the image data. A hybrid system was used, which effectively involved rasterizing the census tracts, and removing those parts which were interpreted as non-residential land uses in the RS data.

Many subsequent GIS installations have developed along the broad lines of these two major examples, often seeking to integrate data for land management over large areas. In Europe, applications have tended to be some years behind the initial developments in North America, as the agencies concerned are often responsible for smaller areas and working under tighter financial constraints, which may have discouraged the early use of satellite RS data. Also, in the UK, the quality and coverage of existing paper mapping are far in excess of those in most other countries, and hence there has not been the same incentive to use the new technology to complete topographic mapping of extensive remote regions (Rhind, 1986). As software for these applications has become commercially available, more organizations have begun to enter the field. Siderelis (1991) illustrates a number of 'land resource information systems' applications, in which environmental management problems have been addressed using GIS. These include a management plan for a major estuarine environment, site screening for a hazardous waste site, and locating a site for a specialized scientific research facility. In each of these cases, the GIS is used to explore a number of possible future scenarios and to assess their impacts on the environment entirely by modelling within the computer. Each of these applications also has socioeconomic components, in that they are all concerned with the relationship between the physical environment and human society.

The CORINE (COordinated INformation on the European environment) programme provides an example of an international environmental GIS application (Wyatt et al., 1988; Mounsey, 1991). The objective here was to establish an environmental database for the European Union, and contained aims both to rationalize data collection and availability and to develop appropriate techniques for its storage and manipulation. The data considered suitable for inclusion in the system should ideally already be in digital form, and would be restricted to the smaller mapping scales in order to preserve a manageable data volume. The information covers themes such as water resources and quality, atmospheric pollutants, slopes and erosion risks, administrative units, and basic socioeconomic data. The system was implemented using the Arc/Info GIS, and important issues to be tackled concern the generalization of data between the different source data scales. In the end, data were held in two forms, notionally corresponding to map scales of 1:1,000,000 and 1:3,000,000. Particular problems with a system of this kind concern the uneven geographic coverage: information from different countries is temporally and spatially inconsistent, and some coverages are not available across the whole of the EC. The potential for variations in data quality between the many sources is immense. Also there is no clearly defined model for data access: options include a centralized database, as in the pilot study, duplicate copies of the database at a number of different sites, and some form of distributed database. The creation of a major database of this kind from many different sources raises important questions about the extent to which the database designer should seek to protect or warn the user about the quality of the data, and its likely implications for particular analyses. It is common for many of the major issues concerned with GIS establishment to be of an organizational rather than a technical nature: GIS technology cuts across traditional disciplinary and professional divisions, and success-ful data integration requires a high degree of inter-agency (and in this case, international) cooperation.

GIS installations concerned with the management and monitoring of the physical environment may be found at many different scales. The use of GIS for the integration of data concerning many different aspects of the environment allows simulation of the effects management plans, and may have a substantial input to decision-making processes. As these applications have grown, a number of new texts such as Haines-Young and Green (1993) and Goodchild et al. (1993b) have appeared which are concerned with the development of GIS with appropriate data models and function-ality for such uses. An example of such an application in a rural environ-ment would be the assessment of the impact of crop spraying on the water quality in a reservoir, as illustrated in Figure 3.1. An elevation model may be interrogated to determine the extent of the reservoir's catchment area. The volume of rain falling on the catchment, and information about crop types and spraying practices may then be used to calculate the quantities of chemicals reaching the reservoir. (The procedures used for this kind of manipulation and modelling of spatial data are explained in Chapter 7.) Once such models are established, the effect of proposed management changes on the local environment may be rapidly assessed.

BUILT ENVIRONMENT

The second major area of application of GIS has been in monitoring the built environment, in which context a very large number of municipal tasks may be identified (Dangermond and Freedman, 1987; Parrott and Stutz, 1991). These applications have been generally vector-based, as there is a need for a high degree of precision in the location of physical plant and property boundaries (Lam, 1985). There is a very direct link between the systems used in these applications and conventional CAD systems, with many agencies converting from CAD to GIS, or running both systems in the same application area.

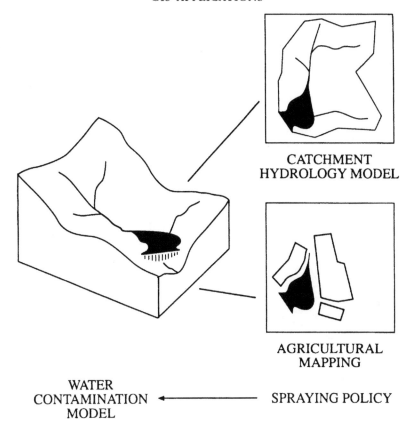

CATCHMENT
HYDROLOGY MODEL

AGRICULTURAL
MAPPING

WATER
CONTAMINATION ←——————— SPRAYING POLICY
MODEL

Figure 3.1 A natural-environment application: modelling water contamination from crop spraying

In the UK, some of the major actors in this field have been the public utilities, with utility applications being evident as a significant driving force in the adoption of large-scale GIS. These organizations, with their extensive use of paper plans to record, for example, the location, diameter, material and installation details of underground pipes and cables, were among the first to realize the potential of digital mapping for efficient database update and retrieval and for data integration between agencies (Hoyland and Goldsworthy, 1986; Mahoney, 1991). Application areas within the utilities include network analysis, fault location, marketing analysis, development planning and emergency planning. Two influential pilot studies undertaken jointly by a variety of local utilities in the UK were the Dudley and Taunton trials (DoE, 1987). In the Dudley experiment, one databank was shared by all the utilities, whereas in Taunton, separate copies of the complete database were installed on each utility's own system, with updates being issued and distributed on magnetic tape. Many utility plans are based on OS base information, and the slow production of a national digital topographic coverage has led to considerable vector digitizing of topographic detail by the utilities themselves, some of which has been done to OS standards. One of the reasons for the selection of Dudley as a pilot area was the existence of detailed OS digital data for that area. An example of in-house digitizing is the use of 1:1,250 OS information by the Welsh Water Authority as background to its own water-distribution database, surveyed in detail by the utility itself (Gunson, 1986; Whitelaw, 1986). In other installations, where integration of background and network information is less important, OS information has been captured

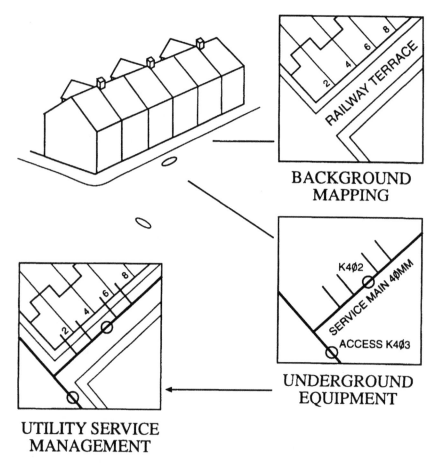

BACKGROUND MAPPING

UNDERGROUND EQUIPMENT

UTILITY SERVICE MANAGEMENT

Figure 3.2 A built-environment application: utility mapping using background digital map data

with a raster scanner (Alper Systems, 1990) to provide a background for utilities' own distribution network information. Such systems allow the preparation of detailed maps and information for teams sent out to undertake roadworks, and where records are shared, costly damage to other organizations' systems may be avoided. Data about local pipes and cables may be rapidly retrieved in response to customer reports. The types of information commonly found in such applications are illustrated in Figure 3.2. Similar applications have been widespread in the USA, with Mahoney (1991) noting considerable variation in the systems which are being developed.

Another area of application relating to the built environment is that of cadastral mapping, where the basic units in the database are ownership parcels. Kevany (1983) outlines the concept of an integrated system based on individual parcels and serving many purposes, but stresses that it remains only a concept. Chrisman and Niemann (1985: 84) also point towards such a system, observing that at present 'systems are purchased by agencies and companies to assist in their pre-established roles'. As organizational practices adapt to the new technology and traditional distinctions become more blurred, the coordinate system, not a base map, must be the basis for the integration of digital records. This places great importance on digital-data quality and data-transfer standards. In the UK, Her Majesty's Land Registry (HMLR) is the organization most directly concerned with land ownership parcels. HMLR has carried out a number of pilot schemes linking existing attribute database systems to digital boundary information (P. Smith, 1988), but is not at present in a position to greatly extend the coverage of this information. The database is at present address-based, with references to postal geography and paper map sheets, but this is being actively extended and developed to create a more powerful digital resource. More widely used in the UK have been address-referenced registers such as the rates register and council tax register, described by Martin *et al.* (1994), which provide limited information about property types and values, but require georeferencing using some intermediate directory or data product before GIS applications are possible.

In Northern Ireland, where the actions of the public utilities and Ordnance Survey (OSNI) are all governed by the Department of the Environment, there has been considerably more coordination than in Great Britain. Brand (1988) describes the implementation of the topographic databases for large- and small-scale mapping on a Syscan 'turnkey' system, with additional attribute information in an associated Datatrieve database management system. The utilities have been involved in various pilot projects using both OSNI's spatial data and their own attribute databases, working towards a truly integrated GIS for the Province.

HANDLING SOCIOECONOMIC DATA

It is now necessary to look in some detail at the automated systems which have been used to handle data relating to the socioeconomic environment. In most applications to date, these data have been derived from censuses, although other sources of information are increasingly becoming available, as the results of surveys and routine registration information are automated (Rhind, 1991). In many cases, the data relate to a variety of basic spatial units, according to the purpose of the data-collection exercise (e.g. census enumeration districts, travel-to-work areas, unit postcodes, store catchment areas, etc.). This variety of basic units has presented a major obstacle to the analysis of socioeconomic data, as the patterns apparent in the data may be as much due to the nature of the collection units as to the underlying phenomena, and there is no direct way of comparing data collected for differing sets of areal units (Flowerdew and Openshaw, 1987). Many of the systems which have been developed in this field are little more than census mapping packages, which are in many ways similar to vector CAC systems. Although the functionality of these mapping packages has been limited, the data structures required for the construction and recognition of areal units are more sophisticated than those used for simple point and line data layers in CAC. Data structures are considered in more detail in Chapter 6. Even very simple mapping systems which are designed to handle census data require different default values and mapping conventions from those for the presentation of physical environment data. These software systems cannot really be called GIS, according to the definitions adopted here, as they mostly lack the spatial and attribute manipulation abilities which have been used to distinguish between GIS and CAC, but they may include more flexible data models.

In the following section, we shall examine two of the earliest and most influential systems for mapping census data, the American DIME and TIGER systems. On pp. 42–43 we shall consider the very many census mapping packages which have evolved subsequent to DIME, often in the context of specific application areas such as health-care management and epidemiology. The focus of the main discussion is on census-derived and medical data, but, finally, some attention is given to the more general area of 'geodemography'. This represents the relatively recent explosion in the acquisition and analysis of spatially referenced socioeconomic data, which is forming the basis for further development of population-based GIS.

DIME and TIGER

Following experimental mailing in 1960, US census forms have been mailed out to most households, and the majority are also mailed back. Enumerators are therefore required to visit only a small proportion of the enumerated addresses, in order to follow up missing forms. An accurate national address list is essential to the effective organization of a census in this way. In addition, there is pressure for detailed data to be aggregated and made available for ever-smaller geographic areas. This requirement must be set against the absence of a large-scale national mapping series, such as that which exists in the UK. In response to this need, the DIME (Dual Independent Map Encoding) system was developed for the US 1970 Census, initially for metropolitan areas only (Barr, 1993a). The DIME scheme was a street-segment-oriented system for boundary definition which was originally developed for nearly 200 Standard Metropolitan Statistical Areas (SMSAs) (Teng, 1983). The spatial data defining street segments (i.e. the section of street between two adjacent intersections) were stored in a Geographic Base File (GBF), hence the term 'GBF/DIME'. This file formed the basis for many mapping and georeferencing operations. The information recorded for each street segment included the

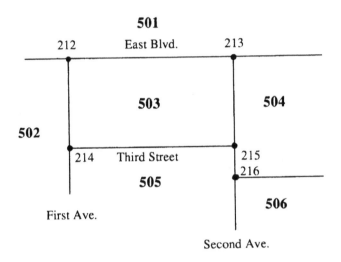

SEGMENT	START NODE	END NODE	LEFT BLOCK	RIGHT BLOCK
Third Street	**214**	**215**	**503**	**505**
.	.	.	.	.
.	.	.	.	.

Figure 3.3 Information contained in a DIME-type geographic base

39

address ranges represented and identifiers for larger geographic units on either side of the segment, such as city block and census tract. The spatial reference was by means of coordinates for the node points at the ends of each straight line segment. Figure 3.3 illustrates an example of the type of information contained within such a base file. Coverage extended to inner-city and contiguous suburban areas, with additional records for significant non-street segments such as railways and rivers. Following the 1980 Census, the coverage was extended to include 278 SMSAs, and 61 other towns for which GBF/DIME files were created. A similar street-based encoding system has been developed for Canada, using the street-intersection node as the basic spatial entity, to which attribute information may be attached.

The segment records in DIME provided a spatial structure to which attribute information could be related. This could be performed either by polygon construction for choropleth mapping by census tract (Hodgson, 1985), or by the use of the other geographic codes associated with each segment for address matching and geocoding. The GBF/DIME structure, with its ability to define many different areal units for which socioeconomic data may be obtained, thus offered a framework for true GIS operations, in addition to simple mapping. For example, when address-referenced records were available, an explicit link existed in the data whereby all area identifiers associated with the matching segment record could be assigned to the street address reference.

The DIME system was extended to cover most urban areas for the 1980 Census, although the data still only covered approximately 5% of the land area, despite including all the principal SMSAs. For the 1990 Census, a more sophisticated system incorporating both data files and associated software has been developed; this is known as 'TIGER' (Topologically Integrated Geographic Encoding and Referencing). TIGER maintains links with 1980 boundaries for evaluation of population change between the two censuses (Kinnear, 1987; McDowell et al., 1987). The TIGER database comprises a hierarchy of files, which include both the DIME-type street-level information and additional levels of area cataloguing which contain definitions of political and statistical areas, both current and historic. It is the intention that TIGER should be maintained as an ongoing national geographic database appropriate for georeferencing of addresses and for use as a sampling frame for population censuses and surveys (Marx, 1990). As its name suggests, the topology of street segments, intersections and enclosed areas is a key feature of TIGER's data structure (Broome and Meixler, 1990). 'Topology' here describes the structure of the street network in terms of links and nodes, identifying which streets meet at each intersection and providing the basis for extracting specific city blocks and tracing complex routes. Information about the shapes of street segments has enhanced the usefulness of the data as a base for digital map production. The database provides national coverage without gaps or overlaps and has formed the basis for many new GIS applications and software products which are not directly related to its original census purpose. US freedom-of-information legislation has ensured that the TIGER data have been made widely available, and a variety of value-added versions of the national database are available (Thrall and Elshaw Thrall, 1993). The existence of a database of this nature offers enormous potential for the development of specialized software by individual agencies, for tasks such as vehicle routeing, market research and direct mail targeting (Cooke, 1989). Despite the inevitable errors and difficulties with such a large database, it provides those in the USA with an invaluable resource for much applied socioeconomic work.

A comparable system has been developed in Canada, in which digital boundary files are nationally available corresponding to the geography of the 1991 Census. Detailed street-network files (previously called 'area master files' or 'AMFs') are available for most large urban centres. As in the USA, the street-network data lend themselves to GIS applications such as route planning, where it is necessary to relate population distribution to other features of the built environment. Again, there is a close link between the census geography and postal codes, and a conversion file exists, which allows the assignment of postcoded addresses into census areas and the provision of corresponding grid references (Statistics Canada, 1992).

Census mapping systems

Unlike the USA with its DIME and TIGER systems, the UK has had no general scheme for the generation of a digital base file from which all significant socioeconomic areal units may be derived. In the UK, the areal units are highly irregular, and are defined in terms of a wide variety of ground features such as field boundaries, roads, railways and streams. Although OS data may potentially contain all the necessary features, problems exist with the extraction of boundaries from many different layers of essentially CAC-type information, and the incomplete nature of the database precludes such an extraction on a national basis anyway. Digitized boundary data created following the 1981 and 1991 UK Censuses have been labelled only in terms of census and some other administrative geographies (health authorities, parliamentary constituencies). It is therefore not possible to directly link these data with that from other areal units without recourse to directories or areal interpolation procedures. More consideration is given to the various indirect methods which have been used for linking records on pp. 118–121. Many agencies have been involved in the creation of digital boundaries for specific localities apart from national data series, and have created base files for smaller areal units in the context of particular studies.

Where access is possible to well-structured spatial boundary data and census information, considerable sophistication can be achieved in the use of GIS-type operations on these data. Examples include the identification of population characteristics falling within specified drive-time zones of a proposed development and the assessment of populations exposed to high levels of aircraft noise around an airport. Applications such as these are dependent on the establishment of geographic relationships between road-network data or noise-level models and the locations of particular units for which population data are obtainable. Using a georeferencing system such as TIGER, advantage may be taken of the explicitly labelled topographic information about roads and other identifiable features, but this is more difficult where areal units are irregular and boundaries are significant only in terms of the data-collection units. A common weakness of such applications is thus their dependence on the size and shape of the data-collection areas and the inability to directly associate areal characteristics with an individual address-referenced record which falls within that area. These general problems with areal population data are examined at some length in Chapter 9.

The paucity of boundary information available in the UK has limited the geographic use of census data for many years to simple choropleth (shaded area) mapping. Until the 1991 Census, no digital boundary data were generally available below the ward level of the census geography (examined in Chapter 5). Agencies interested in the analysis of socioeconomic data such as health authorities, local-authority planning departments and academic institutions have generally not had the resources to develop sophisticated systems such as those seen elsewhere, and the actual data available would not support large-scale geographic manipulation and query.

Applications using census-based data sources have tended to make use of smaller computers and less costly software than the physical-environment applications already considered, due to the relatively straightforward data involved. Digital boundary data and census information are characterized by large storage volumes, but involve unsophisticated geographic processing operations. This also explains the recent emergence of a number of census mapping systems in which national data are stored on CD-ROM and software runs on personal computers. Wiggins (1986) reviewed three low-cost mapping packages for microcomputers, and additional systems have appeared in association with the census data products of the early 1990s. There has also been considerable use of general-purpose mapping systems. PC-based systems have in the past been limited by hardware constraints of memory and disk space, speed and graphics quality (Gardiner and Unwin, 1985), but modern machines offer no real obstacles to the simpler mapping and query operations, even where data volumes are large. Some of the more modern software produced for microcomputers such as Saladin and MapInfo offers significant GIS processing capabilities. It is the issues relating to socioeconomic data encoding and representation which are more enduring and fundamental, and these therefore form an appropriate basis for further research and development. Plate 1 illustrates a shaded-area (choropleth) map constructed from the 1991 UK Census data, and digital boundaries from a national dataset at the enumeration district (ED) level, which is typical of the output from many current census mapping systems. The map shows the percentage of population under 10 years of age for the City of Southampton. The broad pattern of concentration of young families in the centre and towards the peripheral housing estates is apparent, but the detailed geography is obscured by the wide variety in shapes and sizes of the data-collection zones. A number of the larger EDs have very small populations, comprising mainly dockside warehouses (south) or recreational land uses (north centre).

A common application area for these types of mapping system has been in health studies (Gatrell, 1988), and it is worth noting some key features of this work. In the UK, this serves as an example of the existence of very large (potentially) spatially referenced databases held by public sector bodies. The health sector also demonstrates many of the difficulties in making effective use of standard socioeconomic datasets within GIS. These data have been collected for some years, without any particular use being made of the inherent geographic information until very recently. Bickmore and Tulloch (1979: 325) point to the enormous potential of studying disease patterns 'on an ephemeral computer graphics basis rather than by concentrating on producing a massive stack of simple subject maps'. They note that mapping has been used for many years as a tool in medical statistics. It is important to make a distinction between two application areas here: epidemiology and health-care planning. Although the new technology is relevant to both, the methods of implementation and use will be rather different. Initial developments such as the automated systems described by Bickmore and Tulloch were aimed at the analysis of mortality and morbidity patterns. The more modern PC-based mapping systems have been targeted rather more at health care managers and planners as a tool for browsing extensive databases and producing profiles of likely demand for services in specific localities. Gatrell (1988) stresses that statistical analyses are not difficult to program, and one of the most promising avenues for development is the use of geographic databases with specialized medical statistics modules, although this has not generally been done so far. One application of this type is the 'Geographical Analysis Machine' (GAM) developed in the context of the analysis of cancer clusters in northern England (Openshaw et al., 1987). The GAM undertakes a fully automated analysis of a set of point-referenced data, exploring the database for evidence of pattern, without any predetermined information or specification of hypotheses. The technique is only possible with the use of a sophisticated GIS data structure and retrieval mechanism. The point data in this context are derived from a regional cancer registry in which patient

43

addresses are georeferenced by use of the postcode system, another example of the power of GIS to bring together disparate databases for geographic analysis. Clearly, there is potential for the integration of many other datasets held by health authorities. UK examples include Patient Administration Systems (PAS) and Hospital Activity Analysis (HAA) data. Since the publication of the report of the Korner Committee (Korner, 1980), health authorities have been encouraged to postcode records as comprehensively as possible. In this case, we are seeing further growth of geographic data, as geocodes are added to previously existing aspatial records.

Drury (1987) describes a system designed to provide managers with indicators of the existence of gaps and overlaps in services and background information on the characteristics of their areas. He stresses that these should be supplemented by knowledge of the local situation. Such systems provide basic thematic mapping of census-derived data, with some mechanism for linking postcoded records to the census data. This is achieved via the Central Postcode Directory (CPD), described in the following section. The use of either of these systems assumes the existence of a local digital boundary database at the enumeration district (ED) level. Hirschfield et al. (1993) explore a different type of application, in which GP catchment areas are examined in terms of local travel times computed within the GIS using detailed road-network data. This work clearly has relevance to the definition of catchment areas and the measurement of accessibility to local health-care services.

One of the most important influences in this field is the organizational change occurring within the UK's NHS in the early 1990s. Under the new 'purchaser' and 'provider' system, district health authorities are responsible for purchasing the most appropriate health-care services from a variety of centres on behalf of their local population (Hopkins and Maxwell, 1990). In order to inform these decisions, enormous quantities of accurate information about the distribution and needs of the local 'client' population are required, and this has prompted large-scale interest in the potential of GIS as a health care management tool (Wrigley, 1991). This interest in the health sector serves to illustrate the type of application field in which properly formulated socioeconomic GIS may become invaluable to the efficient functioning of organizations. The use of GIS by UK health authorities is surveyed by Gould (1992), who notes a generally high level of awareness of spatial data and GIS capabilities among health-service personnel. However, he concludes that much of contemporary activity is of a low-level nature, typically involving the production and visual inspection of thematic maps, and that there remains enormous potential for the development of geographic analysis in this field.

Mohan and Maguire (1985) illustrate some of the potential for such systems, particularly citing vaccination and immunization data, information on maternity and related services, and details of general-practitioner work-loads as areas which would greatly benefit from greater geographic under-standing. Conventional mapping of census data in the context of evaluating GP workloads has already been a matter of interest in the medical profession (Jarman, 1983). What is required in this context is not fully functional turnkey systems, as have appeared on the market for physical-environment applications, but specialized systems with functions which are relevant to the particular types of data involved. In this case these would include, for example, the calculation of mortality and morbidity rates and the definition of general-practitioner and clinic catchment areas. Initially, these systems have tended to be simple implementations on microcomputers, dealing with aggregate data, primarily for use as a management tool and to demonstrate the capabilities of the technology. The way is now becoming clear for much more sophisticated analytical systems which may be used in an operational context. Mohan and Maguire suggest that 'the merit of GIS is that, by linking datasets at small spatial scale, they offer a reasonable compromise between the accuracy of individual data and the generality of aggregate data'. Again, the existence of national property-level georeferen-cing may have potential for very detailed locality studies in this field, but concerns over data confidentiality and the inadvertent identification of individuals are likely to restrict developments of this kind.

All these systems suffer from a number of weaknesses which relate not to the technology itself but to the nature and quality of the data to which it has been applied. Carstairs and Lowe (1986) review the main issues in the creation of a spatially referenced database for epidemiological analysis. Problems include the variety of areal bases for data collection; the difficulty of relating point- (address-) referenced health-event data to other areal phenomena; the long time interval between censuses; and difficulty in the calculation of morbidity rates, both actual and expected.

Most of these difficulties fundamentally relate to the spatial nature of the data held in the computer. While it may now be possible to transfer a wide range of GIS technology from applications in the physical environment, this will not address the issue of the most appropriate digital representation of the socioeconomic environment. Even in situations where the available data conform more closely to the precise physical model (e.g. DIME/TIGER), the nature of the population-based phenomena and their relationship with the spatial units used has been overlooked. Studies, such as that of Carstairs and Lowe, which address data problems, have adopted an essentially pragmatic rather than a theoretical approach.

Geodemography

In the preceding sections, we have noted the increase in the availability and use of spatially referenced data. The growth of these data and their use in relation to socioeconomic phenomena has become known as 'geodemography' (Beaumont, 1989). Many organizations, including health authorities, retailers and direct mail agencies, have become very interested both in the description of geographic locations in terms of their socioeconomic characteristics and the identification of localities containing people of specific socioeconomic profiles (e.g. poor health, high disposable income, etc.). It is this combination of spatial and attribute queries on essentially the same database which seems to suggest that GIS would be an ideal framework within which to integrate such data. We will here consider a few of the developments which are taking place in this field, as they clearly have relevance to the construction of a GIS to handle population-based information. Brown (1991) notes that the whole field of creating neighbourhood typologies has long interested social geographers, but that the emergence of geodemography as a multi-million-pound industry has been very recent.

Beaumont (1991) stresses the importance of census and postal geographies which allow the characterization of small areas in terms of their social characteristics and the association of such indicators with specific groups of addresses. Census information is a particularly important input to most geodemographic systems, as it provides the basis on which to perform various multivariate statistical analyses resulting in the identification of a number of distinct neighbourhood 'types'. Consequently, there is major recalculation of the standard geodemographic indicators when the results of a new census become available. Many of the leading systems also incorporate other data sources which are complementary to the census, and provide additional dimensions to the classification which are not available from census data alone, such as income levels or daytime populations. Until the mid-1980s the smallest spatial unit generally available for the analysis of socioeconomic data in the UK was the census ED. Various agencies have attempted to classify EDs according to combinations of census variables, initially using 1971 data (Openshaw et al., 1980; Webber, 1980). From these classifications evolved a variety of more refined methods, variously distributed as ACORN (a classification of residential neighbourhoods), PIN codes, MOSAIC, and Super Profiles. The neighbourhood types identified are not tied to specific statistical thresholds, such as 'above 80% owner-occupied housing', but result from the characteristic groupings of many different input variables. The suppliers tend to label the resulting classes according to the neighbourhoods which they most closely represent, such as 'older couples in leafy suburbs' or 'younger families in low-rise local authority accommodation'. Plate 2 illustrates a CCN's neighbourhood classification scheme, showing the different 'lifestyle groups' for a west London catchment area. As the commercial potential of these techniques has been realized, enhancements have been made by using additional or more up-to-date data from other sources and the customization of areas to provide more meaningful neighbourhood-type regions. Beaumont (1991) identifies the major geodemographic applications as being management information systems, branch location analysis, credit scoring and direct marketing. Comparison is made between the situations in the UK and the USA by Flowerdew and Goldstein (1989).

Increasingly, use has been made of the postcode system as a means of georeferencing non-census data (Hume, 1987). An explanation of the geography of the postcode system is given in the discussion of socio-economic data collection in Chapter 5. In the UK, postcodes are alphanu-meric codes (though more generally numeric codes are used elsewhere), which uniquely identify a small number of street addresses, and using a computer file commonly known as the Central Postcode Directory (CPD) or Postzon file, it is possible to obtain a 100m national grid reference for each postcode. A second file, the Postcode Address File (PAF), contains a list of all addresses which fall within each postcode, although giving no explicit spatial reference. The availability of grid references for postcodes has opened up the potential for identifying postcodes which are geogra-phically coincident with populations displaying specified characteristics. For example, a direct mail organization may know from survey data the socio-economic characteristics of their existing customers. Using one of the commercially available ED classification schemes, it would be possible to identify areas with similar aggregate characteristics, which are likely to contain potential new customers. Geographically linking the ED to post-codes, and postcodes through the PAF, it is possible to prepare a mailing list for promotional literature, targeted at areas containing the most likely customers. A similar principle (Jarman, 1983) is applied to the identification of underprivileged neighbourhoods for the allocation of additional health-care resources. Brown (1991), in a discussion of the linkages between geodemography and GIS, identifies spatial linkage as one of the key issues, and notes that many of the existing operations may be treated as list-matching exercises, without the need for sophisticated GIS software. However, the advent of more detailed georeferencing systems, such as the Pinpoint Address Code (PAC) and Ordnance Survey's ADDRESS-POINT™ (discussed in Chapter 5), makes possible a range of more complex geographic queries, particularly concerning travel times and the detailed geography of local populations. There is also a current trend towards specialization and customization of market indicators, with increas-ing use of market-specific information rather than a simple reliance on the general-purpose classifications obtainable using census data. Openshaw (1995) considers the potential role of more advanced spatial analytical techniques in marketing applications.

One final project worthy of mention here is the BBC Domesday system, developed to present a record of Britain in 1986, the 900th anniversary of the original Domesday Book commissioned by William the Conqueror. Further reference is made to aspects of the database construction in

Chapter 9. Data for the system are held on two CD-ROM video disks, allowing very rapid retrieval from around 400 Mb of digital data (1 Mb = 2^{20} bytes). The project is of interest here because it represents a massive collection of social, economic and cultural information, accessible by geographic enquiry (Owen et al., 1986; Openshaw, 1988). Collation of the data for the system demonstrated the huge variety of areal units at different scales which are in use, and indicated 'the severe difficulties faced by geographers in integrating social and economic data from different sources in exploratory analyses of socioeconomic phenomena' (Owen et al., 1986: 309). The Domesday project, although limited by the use of non-standard hardware and the lack of updating, provides an interesting example of a GIS-like approach to the very widespread dissemination of spatial data.

SUMMARY

We have now considered examples of the types of environment in which GIS technology has been applied, with particular reference to the potential for such systems in socioeconomic applications. For a variety of reasons, including the specific application interests of early system designers, the majority of GIS have been targeted at the processing of data relating to the physical environment, both natural and built. The ability to conceptualize and measure the location of physical objects (e.g. roads, rivers) according to some common coordinate system makes their spatial representation relatively unproblematic. This has led to a substantial commercial market for general-purpose systems, used by utilities, land-use managers, etc. However, more recently, there has been a substantial growth in the collection and use of georeferenced data relating to the socioeconomic environment. Increasingly, accurate geodemographic information is considered to be of considerable commercial value to a wide range of organizations. However, phenomena relating to people (e.g. unemployment, deprivation) are inherently more difficult to represent spatially, as it is not usually possible to define precise locations. Georeferencing is therefore frequently indirect and incompatible, via geographic codes such as census areas or postcodes, although new data products are appearing which offer property-level georeferencing. The imprecise nature of the majority of these data makes them unsuitable for use in conventional GIS data models, and consequently less sophisticated mapping systems and analysis techniques have tended to be applied. The difficulties encountered mostly relate to the mode of representation of the data themselves, and it is to this area that more careful consideration must now be given, especially in the light of the massive potential of good spatial representations of the socioeconomic environment. Although GIS would appear to offer a powerful tool for increased understanding of human-environment interaction, the appropriate use of the available data will always be heavily dependent on the quality of the data models around which systems are constructed.

One aspect of GIS development which has become clear, is that the development of the technology has been fundamentally application-driven and guided by very few theoretical considerations. This is particularly the case with regard to the fundamental properties of the models of the world held within these systems, and is nowhere more evident than in geodemographic applications, in which the characteristics of a population are estimated from a number of incompatible and overlapping aggregate datasets. The best foundation for the extension of GIS technology into different application fields, such as the modelling of socioeconomic environments, is a clear theoretical understanding of the processes at work, and it is to these considerations that we turn in the following chapter.

4

THEORIES OF GIS

OVERVIEW

In the previous chapters, we have seen the way in which, for the greater part of its history, GIS development has tended to follow technological advances both in application fields and in other types of information system. This characteristic of GIS is mirrored in the conceptual and theoretical work which has appeared. These diverse application fields have tended to prevent the emergence of any general understanding of the way in which GIS represent the geographic 'real world' or of any discussion as to whose definition of the 'real' is most appropriate. The existing theoretical work may be characterized as addressing two main themes: (1) the 'components of GIS' and (2) the 'fundamental operations of GIS'. However, when considered at a deeper level, these avenues of thought strongly reflect the technological roots of GIS development and fail to address the wider issues of spatial data handling and the GIS as a spatial data processing system. Goodchild (1987: 327) noted, 'to an outsider GIS research appears as a mass of relatively uncoordinated material with no core theory or organizing principles', and although the research community is now more strongly organized, theoretical work is still relatively rare. At the start of the 1990s, Goodchild (1992: 43) was still able to note that 'few people have had the time to write the textbooks or to identify the intellectual core'. It is a measure of the inadequacy of the existing formulations that they offer no real assistance to the application of GIS techniques to socioeconomic data except as a checklist of possible software functions. This is because the concepts used, being essentially at the level of software description, do not offer any explanation at the more complex level of data representation.

We shall begin by considering the nature of objects existing in geographic space and the way in which they are represented by mapped data. The theoretical outlines for GIS are then reviewed, and some attention given to the question of how best to represent geographic objects in a GIS. Finally, we consider the place of GIS within the more general environment of

50

scientific enquiry. A framework is presented here which encompasses the components and operations views of GIS, while focusing on the transformations undergone by spatial data passing through the system. The framework is based around the introduction of the information system to a theoretical model of the traditional cartographic process, which concentrates on the transformations undergone by spatial data between the real world, collected data, map data and map image. Before addressing these issues in detail, it is necessary to consider carefully the nature of the spatial world which we are seeking to model.

GEOGRAPHIC OBJECTS

In the following discussion, use is made of various terms which need careful definition at the outset. 'Spatial data' is a general term used to refer to measurements which relate to objects existing in space at any scale. These lie along a continuum from the physicist's study of the arrangement of atoms to the astronomer's interest in the pattern of stars in the night sky. Geography is concerned with the study of spatial phenomena from the architectural up to the global scale, and 'geographic data' is the term used to refer to data relating to objects in this range (Abler *et al.*, 1971). This is broadly in accordance with D. Unwin's (1981) range of geographic scales from 100 m^2 up to the size of the earth's surface. Geographic information systems are here assumed to be automated systems for the handling of geographic data. From this type of data GIS should be able to extract intelligence which may usefully be considered as geographic 'information'. These definitions are broadly in accord with two commonly expressed views of geographic information and GIS:

> Geographic information is information which can be related to specific locations on the earth.
>
> (DoE, 1987: 1)

> Geographic information systems are information systems which are based on data referenced by geographic coordinates.
>
> (Curran, 1984: 153)

As described in Chapter 2, the origins of GIS include both the techniques for data handling in general (database management) and those concerned more specifically with the handling of spatially referenced data. This second class operate from sub-geographic (e.g. computer-aided design systems), up to the larger geographic scales of satellite remote-sensing systems and digital image processing. The gradual coming-together of these applications has taken place in response to specific information requirements and largely without reference to any theoretical understanding of the spatial entities represented or the techniques used in their manipulation.

Geography is fundamentally concerned with asking and answering questions about phenomena tied to specific locations on the surface of the earth. A distribution is the frequency with which something occurs in a space. In the words of Abler *et al.* (1971: 60), 'Almost any substantive problem a geographer tackles can fruitfully be considered to be a problem of describing accurately and explaining satisfactorily the spatial structure of a distribution.' This description takes place by means of geographic data, gathered in some way to describe conditions at specific locations and is clearly central to the geographic endeavour. It should be noted here, that much of the language concerned with this type of geographic investigation has its roots in the 'quantitative revolution' which took place in geography in the 1960s and 1970s. There are many contemporary geographers who would perhaps wish to disagree with this definition, and would even dispute the ease with which objects can be meaningfully identified and measured in this way. We shall return to some of these issues later, but, for the purposes of the present discussion, the existence of measurable objects is accepted for the sake of consistency with the cartographic thinking to which we shall refer.

Theoretically, objects existing in geographic space may be described by two types of information, that which relates to their location in space, to which we shall now restrict the term 'spatial data', and that which identifies other properties of the object apart from its location, which we shall term 'attribute data'. Attribute data may be measured according to nominal, ordinal, interval and ratio scales. It is these attributes which are usually the concern of the non-spatial scientist, who uses them to draw up classifications of objects according to the attribute values they possess. The geographer however, is also concerned with classification according to spatial criteria. Dangermond (1983) notes the important quality of variable independence in the representation of such data: attributes can change in character while retaining the same spatial location, or vice versa. A GIS will frequently employ different database management strategies for the handling of these two types of information.

Geographic classification has traditionally involved the subdivision of all objects into one of four spatial object classes, namely points, lines, areas and surfaces. This simple four-way classification has proved useful as an organizing concept for discussion of spatial phenomena (e.g. D. Unwin, 1981), but, so far as the development of spatial theory is concerned, has sometimes been considered a 'shallow' concept which has little to offer (D. Unwin, 1989). The differentiating criterion between these data types is one of dimensionality. Commonly encountered examples of each of the spatial object classes are given in Figure 4.1. Distance (or length) is the fundamental geographic dimension, and spatial objects may be classified accord-

OBJECT CLASS	POINT	LINE	AREA	SURFACE
DIMENSION	0	1	2	3
EXAMPLE	FENCE POST 	ROAD SECTION 	LAND PARCEL 	PHYSICAL TERRAIN 
	+ POST 618B	B2120	PLOT 25	CATCHMENT 3A

Figure 4.1 Examples of spatial objects

ing to the number of length dimensions they possess: zero for an (x, y) point, one for a line, two for an area and three for a surface. Dent (1985) also adds the four-dimensional properties of geographic phenomena existing in space–time, although stressing that geographers are more usually concerned with the first four types of phenomena. Recently, concern has been expressed that GIS do not have adequate mechanisms for dealing with temporal data, although the time dimension is important to many potential applications. These ideas are developed in some detail by Langram (1992), and we may expect to see the handling of temporal data as a future growth area in GIS, although we shall be more concerned with the three geographic dimensions here. Surface phenomena have certain special properties, in that the z value required to define any (x, y, z) location on the surface is actually the value of the attribute at that point. The use of cartographic representations of surfaces has been compared to the use of the ratio scale of measurement of attribute information (Chorley and Haggett, 1965). From a detailed surface model it is possible to directly derive spatial objects of each of the other classes. For example, from a digital elevation model, summit points, watershed lines and areas of land above or below a particular elevation may be obtained simply by examination of the surface values. In a similar way, certain point and line information may be extracted directly from an area data model. The surface model itself cannot be derived from data of any other type without interpolation. There seems then to be a logical ordering of point, line, area and surface object classes such that it is possible to derive lower-dimensioned objects directly from higher, but the reverse is not necessarily possible.

The distinction between spatial and attribute data is a familiar one in vector-based GIS and mapping systems. In addition, the four object classes are the same as those which are frequently used to govern the management of data within a GIS and which form the basic components of any topological data structure (Dangermond, 1983). This is as we might expect from systems which purport to provide digital models of the spatial 'real' world. However, a point which has been largely overlooked both in the geographic literature and in technical writing about GIS is that there is no necessary one-to-one relationship between the spatial objects existing in the real world and the representations of those objects which exist within the data model of the information system.

Openshaw (1984: 3), in a discussion of the modifiable areal unit problem, to which we shall return in Chapter 8, remarks that the usefulness of spatial study depends heavily on the 'nature and intrinsic meaningfulness' of the spatial objects studied. This issue is especially important in the context of GIS, where the nature and meaningfulness of the study depends entirely on the validity of the data model. The GIS user is one step (or more) removed from reality, and their analysis relies on the representation held in the computer. It is therefore necessary for us to consider very carefully the nature of the geographic objects which are represented in GIS, as these are crucial to any subsequent use of the system to answer geographic questions. In the words of Goodchild et al. (1992: 411), 'data models not only define how geographic variation is represented, but also determine the set of processes and analyses that can be undertaken'. Existing theoretical models of GIS have been concerned mainly with description of the internal components of a system and classification of their operations, rather than with the fundamental characteristics of the data on which it operates. It is suggested here that a model of the data environment should form the context within which any attempt to build a conceptual model of GIS should sit.

ANALOG AND DIGITAL MAPS

Geographic data gathered about the environment are conventionally represented in the form of paper maps, which are analog models of the real world. In these 'real' maps (Moellering, 1980), the physical qualities of the lines and areas on the map (e.g. length, thickness, colour) are used to represent certain features of the world. Because the world is too complex to be represented in its entirety, data are selectively measured and stored to produce a scaled-down model of the world (Haggett, 1983). These models are the basis on which geographers answer questions posed about the nature of spatial phenomena. It is therefore crucial that the model be an 'accurate' representation of the world and that it embody the spatial relationships necessary for the solution of any particular problem. What is considered accurate in any particular context will depend as much on our initial assumptions and concepts as on any measurable property of accuracy. A mapping may be thought of as a transformation from one vector space (the real world) into another (the model) (D. Unwin, 1981). The first of these is generally multidimensional and complex, the second is a two-dimensional sheet of paper and the data are generalized. Absolute location in space is conventionally defined in terms of some (x, y) coordinate system which is independent of the objects being mapped. The mapping rules, such as scale and projection, govern the transformation from one space to another. A GIS, in common with an automated mapping system, stores the spatial data required to draw a map instead of a physical copy of the map itself. This model of the world is a digital map.

Certain features of analog and digital maps should be distinguished here. All visual maps display characteristics of projection, scale and symbolization. A large cartographic literature exists which deals with these topics in great detail (e.g. Robinson, 1987). Projection describes the manner in which the spherical surface of the earth is flattened on to the two-dimensional surface of the paper or computer screen. This process will always involve a degree of distortion. Scale is the ratio of distance on the map to the corresponding distance on the ground. On large-scale maps, distortion due to the projection is minimal, and is often ignored. An important issue arising from map scale is the degree of generalization required, as only a small proportion of objects identifiable in the real world can be reproduced on the map, and this proportion becomes smaller as the physical area covered by the map increases. The third aspect, symbolization, describes the graphic symbols used in the map to represent particular phenomena in the real world, such as length, thickness or colour, mentioned above. In an analog map, all these properties are fixed at the time of map production. A digital map, however, contains all the spatial data required for map construction without any of these properties being fixed. Only when it is necessary to produce a graphic image of the map must these parameters be specified. Thus the digital map is amenable to reprojection, scale change and symbolization changes, by mathematical manipulation, and even the content of the visual map may be altered by selective extraction from the digital data. No map is therefore an objective representation of the world, but represents a series of value judgements concerning its content, design and message.

To illustrate this idea, consider a road represented in both analog and digital map form. In the analog version, it is depicted at a fixed scale and projection by a standard symbol, for example a red line, whose thickness indicates its width. This representation cannot be altered until the map is resurveyed and reprinted, and the only information which the map reader can derive is that which is directly measurable from the visual image (such as the distance between two intersections). In its digital form, the road is represented by a series of coordinates, and attribute information about its width and name. This information may simply be plotted according to some convention to reproduce the analog map (which is the role of CAC), but, more importantly, the coordinates may be reprojected, or drawn with different scale and symbolization. A GIS would be able to use the digital information to calculate the exact distance between the intersections (subject to the accuracy of the input data) without recourse to a graphic image at all. Additionally, many fields of attribute information may be held which could not all be shown in a single visual map, but any of which may be included in database manipulation and query.

In contrast to the analog model, it is the geographic data that are the basis of the GIS representation, and not the graphic image itself. Indeed, there may be applications in which the visual image is not necessary, as the answer to a question may be derived directly from the digital database (Cooke, 1989). In both models, 'the principal requirement in cartographic representation is the spatial conformity of the qualitative and quantitative parameters of objects and phenomena to their actual distribution' (Shiryaev, 1987). In some CAC systems, the capacity for these changes is very limited, but more advanced GIS offer highly sophisticated manipulation options. The graphical aspect, so important to the effective communication of spatial information, forms the actual representation of the data in the analog model, whereas the digital map allows for the separation of mathematical and graphical aspects. Despite this greater flexibility, it is important to remember that digital maps frequently begin with one or more analog maps which are then encoded (digitized), and the resulting representation cannot be of greater accuracy or precision than the source data.

THEORETICAL MODELS OF GIS

Reference should now be made to the existing models of GIS operation, which are broadly similar in nature. These may be considered in two main groups: (1) the functions of GIS, and (2) the fundamental operations of GIS. It will be seen that these approaches generally make little or no reference to the very important processes which have been outlined above. It is because of their failure to address these issues that they are unable to offer much help to those who would seek to apply GIS in new situations. In particular, certain socioeconomic phenomena need special treatment in terms of the structures used for their representation in digital systems, but these conceptual models tend only to address aspects of the operation or composition of GIS systems. They make no statement about the nature of the data representation, but the assumption is implicit that the digital map is an accurate picture of the real world. The components in these models are basically analogous to the main software components in any general-purpose system. Bracken and Webster (1989b) note that attempts to classify GIS have typically focused on the task-orientation of systems, reflecting the ad hoc nature of their development. They suggest an alternative classification which recognizes three major components in its characterization of GIS: the problem-processor model, database model and interface model. However, this is still an explicitly software-oriented approach to understanding GIS.

Components of GIS

The basic form of these GIS models is as in Figure 4.2. This model has four major components, and appears in very similar form in a wide variety of papers with variously three, four, five or more main components which are roughly synonymous with the ones illustrated, based on Young (1986); but see also Curran (1984) or Rhind and Green (1988). The key components are as follows:

1 *Collection, input and correction* are the operations concerned with receiving data into the system, including manual digitizing, scanning, keyboard entry of attribute information, and online retrieval from other database systems. It is at this stage that a digital map is first constructed. The digital representation can never be of higher accuracy than the input data, although the mechanisms for its handling will frequently be capable of greater precision than that achieved during data collection.

2 *Storage and retrieval* mechanisms include the control of physical storage of the data in memory, disk or tape, and mechanisms for its retrieval to serve the needs of the other three components. In a disaggregate GIS (Webster, 1988), this data storage may be physically remote from the rest of the system, and may meet the database requirements of other, non-geographic, data-processing systems. This module includes the software structures used to organize spatial data into models of geographic reality.

3 *Manipulation and analysis* represents the whole spectrum of techniques available for the transformation of the digital model by mathematical means. These are the core of a GIS, and are the features which distinguish GIS from CAC. A library of data-processing algorithms is available for the transformation of spatial data, and the results of these manipulations may be added to the digital database and incorporated in new visual maps. Using these techniques it is possible to deliberately change the characteristics of the data representation in order to meet theoretical requirements. It is

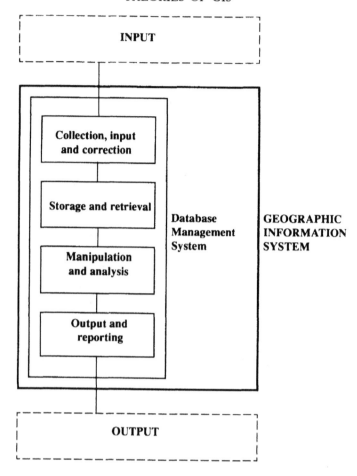

Figure 4.2 A typical model of GIS as 'components'

equally possible to mishandle or unintentionally distort the digital map at this stage.

4 *Output and reporting* involves the export of data from the system in computer- or human-readable form. It is at this stage that the user of a digital map database is able to selectively create a new analog map product by assigning symbology to the objects in the data model. The techniques involved here include many of those of conventional cartography, which seeks to maximize the amount of information communicated from the map maker to the map reader. The nature of the digital model at this stage will have a major impact on the usefulness of any output created.

Fundamental operations of GIS

This approach considers the functions which GIS are able to perform rather than the modules in which these operations conventionally sit. The operations discussed in this section fall entirely within the manipulation-and-analysis component given above, and are thus wholly internal to the GIS. The fundamental classes of operations performed by a GIS have been characterized as a 'map algebra' (Tomlin and Berry, 1979; Berry, 1982, 1987; Tomlin, 1991) in which context primitive operations of map analysis can be seen as analogous to traditional mathematical operations. A distinction is then made based on the processing transformation being performed. These 'classes of analytical operations' are divided into reclassification, overlay, distance/connectivity measurement and neighbourhood characterization, which are, interestingly, independent of raster and vector representations of the data:

1 *Reclassification operations* transform the attribute information associated with a single map coverage. This may be thought of as a simple 'recolouring' of features in the map. For example, a map of population densities may be reclassified into classes such as 'sparsely populated' or 'overcrowded' without reference to any other data.

2 *Overlay operations* involve the combination of two or more maps according to Boolean conditions (e.g. 'if A is greater than B and A is less than C'), and may result in the delineation of new boundaries. In such cases it is therefore essential that the spatial and attribute data are a correct representation of the real-world phenomena. An example would be the overlay of an enterprise zone on to a base of census wards. This would be appropriate to determining the ward composition of the zone, but may not allow an accurate estimate of the population falling within it, as there may not be exact coincidence of the boundaries. Thus the operation is only appropriate if the intended interpretation of the data is meaningful.

3 *Distance and connectivity measurement* include both simple measures of interpoint distance and more complex operations such as the construction of zones of increasing transport cost away from specified locations. Some systems will include sophisticated networking functions tied to the geographic database. Connectivity operations include, for example, viewshed analysis which involves the computation of intervisibility between locations in the database.

4 *Neighbourhood characterization* involves ascribing values to a location according to the characteristics of the surrounding region. Such operations may involve both summary and mean measures of a variable, and include smoothing and enhancement filters. These techniques are directly

analogous to contextual image-classification techniques to be found in image-processing systems.

Sequences of such manipulation operations have become known as 'cartographic modelling'. Berry (1987: 122) notes that 'each primitive operation may be regarded as an independent tool limited only by the general thematic and spatial characteristics of the data'. No further reference is made to the way in which the suitability of these general thematic and spatial characteristics should be evaluated. However, it is suggested here that these 'general thematic and spatial characteristics' are equally fundamental to the valid use of GIS, and that they should form part of any attempt to conceptualize GIS. The nature of the data representation and its relationship to the real world literally define the limits within which such techniques can operate, and it is therefore vital that the concepts of manipulation and analysis be set within a broader understanding of the entities involved. Overlay operations, in particular, draw upon a number of map coverages and may result in the delineation of new boundaries, hence the nature of the data being processed is highly significant. The transformations which data have already undergone at collection and input determine their suitability for these subsequent manipulation operations.

A THEORETICAL FRAMEWORK FOR GIS

This discussion is based on an analysis of the way in which data are transformed and held as a digital model of the external world. The geographic data-processing system outlined is not intended to be a description of any specific software system, but is a model of the processes which may operate on digital geographic data. The idea of data representation used here should not be confused with work on specific data structures, either spatial (e.g. vector, raster, triangulated irregular network, quadtree) or attribute (e.g. hierarchical, relational), which are essentially technical issues and will be addressed in later chapters. Actual software systems may be identified which perform all or some of the principal transformations to a greater or lesser degree, but only those containing some capacity to input, manipulate and output digital spatial data in some form are considered to be 'GIS' in the present context.

The groundwork for this approach has been laid by cartographers seeking to understand the relationships between the world and the map as a model of the world (e.g. Muehrcke, 1969; Robinson and Petchenik, 1975). We have seen how the digital map is related to the analog map, and we must now modify and extend the existing theoretical structure to replace the paper map with a digital map sitting within a GIS. The process of analog map production may be modelled as a series of transformations between the real world, raw data, the map and the map image (Figure 4.3).

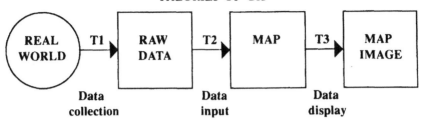

Figure 4.3 Transformation stages in the traditional cartographic process

This approach is echoed in much more recent work on the roles of CAC and GIS (see, for example, Visvalingham, 1989). The significance of these transformations is that they control the amount of information transmitted from one stage to the next. The cartographer's task is to devise the very best approximation to an 'ideal' transformation involving a minimum of information loss. A clear explanation of the 'transformational' view of digital cartography is given in Clarke (1990).

In the context of GIS, we may add an additional transformation stage which sits entirely within the GIS (Figure 4.4). The sequence of transformation stages illustrated in the figure form the basis for the discussion of GIS techniques later in this book. In the first transformation (T_1), data are selected from the real world, as, for example, surveying measurements or census data; these are then input to the GIS in some form (T_2) to provide

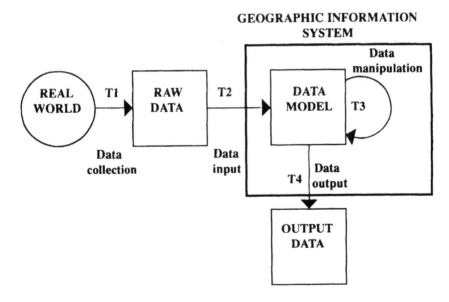

Figure 4.4 A transformation-based view of GIS operation

the basis for its digital map representation of the real world. Within the system, a vast range of manipulation operations are available to further transform the data and store the results (T_3), and these may be communicated as tabular or graphic images via hardcopy or screen (T_4). It is worth noting that each of these transformation stages may actually involve several physical operations on the data, for example, T_1 may involve both collection and aggregation and T_3 will almost always consist of a whole series of data-processing operations. If the 'thematic and spatial' characteristics noted by Berry (1987) are to be understood, it is necessary for us to carefully consider the way in which these four transformations may affect the digital representation of real-world objects.

In the case of data relating to point, line and area objects, the non-spatial attribute information associated with the objects will often remain independent of the transformation operation, but in the case of surfaces, where the attributes are more integrally related to the spatial data, the attribute information may also undergo change. Although attribute values in the database will not be altered by the transformation of points, lines and areas, the interpretation of these values may be affected. Another issue of significance is the fact that the spatial data collected may be only a sample of the set of objects of that class existing in the real world. For example, census data collected for individuals but aggregated and represented as areas present a major problem to interpretation, and cannot be treated in the same way as areal data such as those concerning land use which are both collected and represented as areas. Gatrell (1991) uses the transformation of objects in order to illustrate the types of geographic processing which may be performed within GIS. These transformations may be the inevitable result of some aspect of data processing or may be deliberately brought about by a data-manipulation algorithm. We may therefore identify three possible types of transformation. Each transformation stage in Figure 4.4 may involve one or more of these:

1 those which are essentially geometric, for example changes in the coordinates representing an object's location which are brought about by reprojection of the map;
2 those which affect the attribute characteristics of an entry in the database, such as the reclassification of land-use parcels without alteration of their boundaries;
3 those which are object-class transformations, such as the aggregation of point events to areal units.

At each transformation stage, it is likely that the data will undergo some change as the intended result of the processing operation and some additional change due to the unintentional introduction of error. The act of selecting which information should be included within the dataset may contain components of either of these. The difficulties of avoiding data-input T_2 errors are well known (these are examined in some detail in Chapter 5). Newcomer and Szajgin (1984: 59) analyse the accumulation of errors during overlay analysis (a T_3 transformation), and observe that 'map accuracy can be related to the probability of finding what is portrayed on the map to be true in the field'. Others, such as Goodchild and Dubuc (1987) and Veregin (1989), have attempted to model the behaviour of this error propagation. The great danger here is the cumulative nature of errors in a digital database: if the results of one overlay operation result in the creation of a new set of zones containing errors in boundary location or attribute values, these errors will be carried forward into all subsequent transformations using the data for those zones.

The potential differences between the characteristics of the data representation and the corresponding real-world phenomena may be traced to the transformations introduced above. Not only may data be transformed in some way within a spatial object class (e.g. line generalization due to a change in map scale), but the representations of objects may be transformed between classes. Most data input to a GIS are in some way 'secondary' data, in the sense that the act of data collection (T_1) precedes and is distinct from input to the system. These data are then encoded in some way in the process of digitization (T_2). Coppock and Anderson (1987) note the way in which system designers have tended to take for granted the quality and validity of the data collection process. An interesting special case exists where these transformations take place within a system concurrently, such as in the creation of remotely sensed images compiled from satellite scanner data or the input of coordinates directly from a global positioning system. In this situation, the data collection and input operations are combined. In both physical and social survey techniques, recent developments in recording instruments have enabled data entry to be performed by the surveyor in the field, and subsequently uploaded into a central database, again extending the realm of digital data.

Any representation which is of a different spatial-object class from the corresponding real-world phenomenon demands careful attention. Some objects may be considered as belonging to different classes at different scales or in different applications. For example, a road may be treated as a line feature in a regional transportation model, but as an area in a city-scale digital database, on which it is necessary to precisely identify the locations of other, within-street, objects such as lamp posts and traffic islands. (This difference in the way in which objects are viewed at different scales is one of the main obstacles to the creation of truly scale-free digital maps.) Different object-class representations of an object will not yield the same results when passed through one of the GIS transformation stages. This issue is central to the representation of socioeconomic phenomena, as they may be variously considered as point (address), area (administrative zone) or density-surface phenomena, and each of these has very different implications for the handling of such data within GIS.

These observations, if applied to analog map production, give some guide as to the correct interpretation of the map product as a representation of the real world. Not only the end state, or the original object class, but the nature of the transformations must be understood. In the context of a GIS, however, which offers the potential to dynamically manipulate the objects of which the paper map is merely a static model, the situation becomes more complex, and these relationships should be considered as a rule set which determines whether or not a particular manipulation operation is valid, and should be allowed to take place. Such a rule set has considerable relevance to the development of expert-system interfaces to geographic databases, governing the manipulation options open to the user (Chen, 1986). The framework outlined here in no way replaces existing knowledge of the behaviour of spatial objects, e.g. the discussion of the modifiable areal unit problem (considered in Chapter 8), but encompasses these issues, seeking to identify the more general processes at work.

Many data, especially census and other socioeconomic types, are conventionally collected and handled as relating to areal units. The aggregate attribute data are then associated with the boundaries of these areas in the digital map. Consequently the data from individuals are transformed into an area-class geographic database. These decisions are generally made prior to any consideration of the use of GIS. Geographers have long understood the difficulties of dealing with this kind of representation, and the misleading characteristics of choropleth (shaded area) mapping of populations, and the implications of this knowledge for the display and interrogation of data using GIS, are discussed in Chapter 8.

THE REPRESENTATION OF GEOGRAPHIC DATA

We shall now consider the implications of this structure in more detail by following a number of typical GIS datasets (both physical and socioeconomic) through the range of data transformations. An attempt to assess the statistical implications of the transformations undergone by such data is given in Arbia (1989), who also examines a number of additional examples. The examples discussed here are illustrated in Figure 4.5.

A well-known distinction in areal data is that between data collected for 'artificial' and 'natural' areal units (Harvey, 1969: 351). A census district is an artificial areal unit (as mentioned above): the determination of its boundaries bears little or no relationship to the point scatter of individual observations from which its attribute information are aggregated. By contrast, the boundaries of a field are a natural areal unit. The phenomenon 'field' is completely and precisely encompassed by those boundaries,

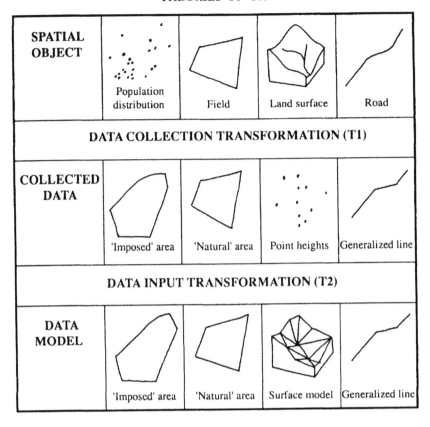

SPATIAL OBJECT	Population distribution	Field	Land surface	Road
DATA COLLECTION TRANSFORMATION (T1)				
COLLECTED DATA	'Imposed' area	'Natural' area	Point heights	Generalized line
DATA INPUT TRANSFORMATION (T2)				
DATA MODEL	'Imposed' area	'Natural' area	Surface model	Generalized line

Figure 4.5 Examples of object-class transformations

and all points within the field share in the common field attributes of ownership, land use, etc.

In the context of census data, the population enumerated (transformation T_1) changes from being referenced as real-world points to area data, whereas the field in the example above retains its real areal classification. If both these areal units are then digitized and input to a GIS, their areal status is maintained through the data input transformation, T_2. Severe limits are imposed on the nature of the manipulations T_3 which we may perform on the census-district representation, as we can make no assertions about the point data from which it was derived. The field data, however, may be treated as a coherent whole, and is safe to process through polygon-overlay or point-in-polygon operations, using any of the attributes possessed by the real area it represents. The significant difference between these two area representations within the GIS is the aggregation operation which took place at the data-collection stage. Because the system representation of the population-based data for the census district is no longer of the same spatial-object class as the real-world phenomenon it represents, both mapping and manipulation operations are severely restricted. The district identifier is actually an item of area-attribute data, not relating to the underlying point pattern, and may be used with validity, in the same way as area data relating to the field.

As a second example, we shall consider data relating to the elevation of the landscape. This spatial 'object' is a true surface, whose characteristics may be measured in a number of ways. In this case, the attribute data are integrally involved in the transformation operations. If data are collected as a series of spot heights (a spatial sample from the infinite number of points existing on the surface), these may be encoded by the interpolation of a series of contours to represent the terrain surface. These line features enclose areas of land between given altitudes. If contours are digitized, we may consider the basic representation within the system to be of line type. Alternatively, the spot heights may be input and used for the construction of a digital elevation model (DEM). This model, although consisting of key data points and interpolated information, is manipulated as a surface, and thus represents the closest which an information system may come to a true surface representation. The DEM offers the greatest potential for the manipulation of the landscape model and is to be preferred over alternative representations such as the contour line model, which may involve generalization and loss of data (Yoeli, 1986). The original data-collection transformation involved the recording of the surface as a series of points, and thus the full detail of the surface can never be recovered, a feature which must be considered in any subsequent interpretation of the results of analysis; but, given this restriction, the best approximation to the real-world data is the data representation which is of the same spatial-object class as the original. This will often involve interpolation to achieve an object class of higher dimensionality, e.g. (surface–point–surface) or (surface–line–surface) in the DEM example.

A third example would be the representation of a pipe or cable distribution network, where there is a direct one-to-one relationship between the points and lines representing inspection covers and pipes and the physical phenomena which they describe. In this context, a linear feature is recorded as a series of coordinates and held in the system by those coordinates. No spatial-object-class transformation takes place at any stage. Information such as the location at which two pipes intersect, or the pipes which may be accessed from a given inspection chamber, may meaningfully be extracted from the database, and is subject only to the locational accuracy of the surveyed data. It is important to note here that these relationships only hold at a given scale of representation: a manhole cover represented by a point in a utility-management GIS may be a whole coverage in a computer-aided design (CAD) system, while the entire utility area may be only one polygon on a national map of utility areas.

Some conclusions may be drawn from these observations: first, that the initial data-collection transformation imposes severe restrictions on what may subsequently be achieved with the data. However, the collected data may be used to model a data representation of the spatial-object class considered appropriate for the original phenomena, and this will provide the most desirable representation option in a manipulation context. Secondly, transformation operations within the GIS may be used to transform data between spatial-object classes, such as the creation of a contour map from the DEM representation and, with less success, the interpolation of a DEM from a contour representation. In all these discussions, a fundamental factor is that of spatial scale. It may be argued that the 'correct' object class for the representation of a phenomenon is scale-dependent. Analysis of geographic data entails assumptions about their object class (e.g. the commonly held concepts of central business districts as areas; roads as lines, etc.). However, these assumptions are only appropriate at a given scale of analysis. If this scale of analysis changes (e.g. to the consideration of central business districts as discrete points in a national study), or the data structure is inappropriate for that scale, the data will not be able to support the queries which are asked of it. For example, a database suitable for the study of unemployment at a regional level will be of little use in the analysis of individual-level unemployment experience. Holding an appropriate model of the geographic world is fundamental to any form of GIS-based analysis. Goodchild (1992: 33) stresses that there are many different possible models which may be adopted for a given phenomenon and the choice between them is 'one of the more fundamental issues of spatial data handling'.

In many situations, including all those illustrated, the important decision as to the nature of the data-collection transformation is made by agencies other than the GIS-using agency. Thus the ideal of data which are of the same spatial-object class as the real-world object is not always achievable. In the case of an elevation model there are simply too many points on the surface for complete enumeration, and any data collected for a real-world surface must therefore be a sample or generalization of some type.

'GEOGRAPHIC INFORMATION SCIENCE'

So far in this chapter, we have concentrated largely on what goes on *within* geographic information systems, but increasingly there is a need to understand the place of GIS in the world *beyond*. We can perhaps most usefully think of this as what Goodchild (1991b, 1992) calls 'geographic information science'. This debate concerns the extent to which GIS is merely a toolbox for the answering of questions with a spatial dimension and the extent to which GIS as a field contains a legitimate set of scientific questions. In particular, there have been lively academic exchanges regarding the role of *geographic* information systems in the discipline of geography (Openshaw, 1991a; Taylor and Overton, 1991).

Most commentators would agree that the study and practice of GIS in itself is neither a science nor an academic discipline, yet it crosses the boundaries of many conventionally defined disciplines. The past tendency for developments in GIS to be driven, or at least limited, by changes in technology and the inevitable advertising hype associated with an emerging software industry, have done little to strengthen its claims to scientific rigour in the eyes of many geographers. The fact that we now have the ability to handle vast quantities of geographic information and to build complex spatial models does not lead directly to the solution of important geographic problems. Taylor and Overton (1991: 1088) outline their 'first law of geographic information' that 'where the need for information is greatest, the amount of information is least'. In other words, the availability of detailed spatial information tends to reflect the interests and power structures which are already dominant in society, and no amount of modelling and analysis of these data will permit us to challenge the existing structure. There is a sense in which this reflects and extends our earlier discussion that our work is fundamentally limited by the nature of the collected data at transformation T_1, i.e. this is not just a technical issue, but a social one also. Much of the disagreement which has occurred can be traced back to exaggerated claims for the power of GIS in the first place, and the challenges facing the GIS user are in many ways the same as those facing the non-automated investigative geographer.

Laurini and Thompson (1992) recognize that their understanding of GIS draws on many different disciplines and fields, including those which provide concepts for dealing with space, instruments for obtaining spatial data, theories dealing with automation, application fields, and those which provide guidance about information. They adopt the umbrella term 'geomatics' to cover all of these, and they consider that aspects of each are necessary in order to formulate an understanding of spatial information systems. This view accepts that there is much to be learned from non-automated spatial information systems and takes a realistic view of what will be possible. They see the potential benefits in terms of faster processing of larger quantities of information, and thus the ability to tackle problems which might otherwise be considered insoluble:

> As a set of software processing routines in a hardware setting they are a new kind of toolbox for practical problem solving. As a new resource compared with paper based map making, they represent a new technology; and, through their emphasis on spatial data, they stand, for

many people, for a different approach to thinking about problems and knowledge.

(Laurini and Thompson, 1992: 24)

Taking this as an indication of what may be expected, we should consider one further aspect that is particularly important to Goodchild's geographic information science, and that is the role of spatial analysis. Following the reasoning outlined above, one of the greatest potential benefits of GIS is the application of a powerful data-handling technology to real-world problems which have previously been hampered by the quantity of information or the time taken in processing. This kind of problem implies the need for spatial *analysis* rather than simple data retrieval or calculation. Undoubtedly many successful GIS installations serve the primary role of data inventories which are useful for day-to-day management of organizations, but surely it should be possible to put all this information and processing power to better use than that? Fotheringham and Rogerson (1993) perceive the delayed development of sophisticated spatial-analytical functions as one of the factors hindering the more widespread interdisciplinary adoption of GIS. Goodchild (1991b) suggests that this has much to do with the relatively unsophisticated market for the major systems, which is concerned with organizational management. There has thus been little incentive for the development and inclusion of complex statistical and analytical functions within commercial software. Some functions are being added to existing software, but many developments are taking the form of specially written tools for particular analyses, either as separate programs, or using the macro programming languages within the larger GIS packages. Many of these techniques originated in geography's 'quantitative revolution' in the late 1960s, but could not be practically implemented for lack of the suitable software environment that GIS should be able to provide. Conceptually, the correct place for such functionality is around the manipulation transformation T_3, which we have already identified as the 'core' of GIS operation. Sui (1994) goes further, and suggests that the implementation of discredited quantitative tools is not the way forward, but that new approaches utilizing fuzzy logic and relative spaces may be developed within the framework of the existing technology.

Fotheringham and Rogerson (1993) identify a number of specific impediments to the implementation of better spatial analysis in GIS, including the modifiable areal unit problem, aggregate versus disaggregate models, and spatial interpolation and sampling procedures, each of which is considered elsewhere in this book. A very important distinction emerges in this discussion, and that is between spatial and aspatial analysis: spatial analysis is not simply the application of aspatial statistical techniques to spatial data. The very fact that the GIS is able to encode both locations and attributes makes possible the development of techniques which incorporate explicitly spatial concepts such as adjacency, contiguity and distance. Anselin *et al.*

(1993) discuss the special nature of spatial data and illustrate differing degrees of integration between spatial analytical software and GIS. Again, the importance of the data model is apparent in determining the kinds of analysis which may actually be supported (Goodchild *et al.*, 1992). The further development of such functions seems crucial to the future of GIS and its acceptability to the academic community: there are genuine scientific questions to be addressed here, and there is an appropriate role for GIS to serve as a toolbox for many disciplines, but there is a long way to go from the practical problem solving to the different approach to thinking.

SUMMARY

In this chapter, a review has been presented of the way in which data are collected and organized about objects existing in the geographic 'real world'. These are the data used to create digital maps – the basis of GIS and CAC. The data stored in these systems serve as a dynamic model of the world, which may be updated, queried, manipulated and extracted by many different processing operations, depending on the application. Existing theoretical models of GIS have tended only to address the functions and component parts of these systems, without reference to the underlying data model – a feature reflecting their often ad hoc and technology-led evolution. It is essential to understand the nature of these data representations and the transformations which they may undergo, and a framework for this has been given. A distinction is drawn at each stage between the operations affecting the geometric, attribute and spatial-object-class characteristics of the data. This formulation gives appropriate importance to the data-collection operations (T_1), and to the nature of the digital model of the world, on which all other system capabilities finally rest. Spatial-object-class transformations are shown to be of major importance to the accurate representation of spatial form and therefore of relevance to all spatial manipulation operations.

These issues are of particular relevance to the handling by GIS of socioeconomic data, which frequently undergoes important object-class transformations at the data-collection stage which have severe consequences for any analyses performed later on. With reference only to the existing conceptual models of GIS it would be tempting to examine the technical aspects of GIS, considered in the following chapters, and to jump straight to the direct transfer of many techniques from physical to socioeconomic applications. However, the framework given here directs our attention not only to the data-manipulation capabilities of GIS but also to the nature of the digital model of the world and the transformations which have been performed in its assembly. We shall return to the implications of this approach in Chapter 9 and attempt to form a view as to the best way of structuring GIS installations for socioeconomic applications.

5

DATA COLLECTION AND INPUT

OVERVIEW

In the previous chapter, a conceptual framework for GIS was presented, in which four transformation stages were identified. GIS provide a mechanism by which geographic data from multiple sources may be built into a digital model of the geographic world. This model may be manipulated and analysed to extract many types of geographic information. The first two of these transformations are data collection (T_1) and input (T_2), shown in Figure 4.4. Many GIS commentators have tended to ignore or underestimate the influence of data collection on subsequent operations, and begin with discussion of data input methods. We have seen how the transformations which data undergo may fundamentally affect their usefulness at later stages, and for this reason, collection and input are considered together in this chapter. Recent interest in the sources of error in spatial data have focused attention more clearly on the methods used for collection and input.

Particular emphasis will be given to sources of socioeconomic information and the ways in which they represent the actual objects of study - the individuals which make up the population. Major data sources include censuses of population, social surveys, and address-referenced information about individuals collected by service-providing organizations (e.g. local government, health authorities, retailers, etc.). Each of these may be spatially referenced in different ways, and the data collection method thus has a major impact on the types of spatial analysis which the data 'model' will support.

These data may be input to a GIS via various different routes: either directly from the collected data by digitizing or by importing the data from some other computer information system. The whole issue of data transfer between GIS installations and data standards is a very important one, which has relevance both to input to and output from systems. It is addressed in Chapter 8, in the context of data output. Direct input by digitizing or by using other data which have not previously formed part of a GIS data

71

model (such as the retrieval of census data from a central data archive) are covered here. These operations may be seen as clearly falling into transformation T_2 identified in Chapter 4. Digitizing frequently involves the use of manual techniques which are time-consuming and error-prone, creating a demand for sophisticated data-verification procedures. These are considered on pp. 84–87. Once data entry is complete, a digital model is available for manipulation (T_3) and display (T_4). The techniques involved in these transformations are examined in Chapters 7 and 8.

One other issue is worthy of mention here: a commonly encountered distinction in GIS is that between vector (coordinate-based) and raster (cell-based) data structures. The whole area of data structures is considered in detail in Chapter 6. The vector/raster distinction has been made much of in the literature, but really only refers to different strategies for modelling geographic reality. Vector and raster representations of the same irregular shape are illustrated in Figure 5.1. These rather crude representations may be made to look more 'realistic' either by using more points (vector) or more cells (raster) to improve the resolution of the image. Some software systems are restricted to the use of vector-only or raster-only data, although all the larger systems have vector and raster modules, with facilities for the conversion of data between the two types. Software which offers true integration between data in the two different formats is scarce. Although different algorithms are necessary for the handling of these two types of data, the underlying transformation processes are the same, and

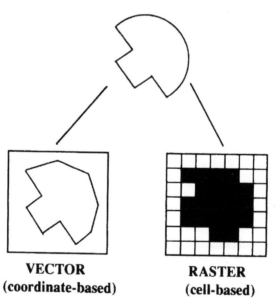

VECTOR
(coordinate-based)

RASTER
(cell-based)

Figure 5.1 Vector and raster representations of an object

both may be used to represent point-, line-, area- or surface-based information.

DATA COLLECTION

Socioeconomic data are collected for a wide variety of purposes and are rarely directly georeferenced (i.e. individual data items do not usually carry a grid reference). Consequently, much of the data is specialized in nature and may not be ideally suited to the requirements of other organizations. In addition, the geocodes used (address references or some form of areal unit) are frequently incompatible with those on other datasets and are liable to change over time. Although geographic location may offer the only means for relating such data, the link is imprecise and often difficult to achieve. The ecological fallacy and modifiable areal unit problem present difficulties when our interest lies with individuals and we are forced to work with data aggregated to areas (see Chapter 8) which must also be considered in this context. To date, few socioeconomic data have been collected with the use of GIS in mind. A clear understanding of the purpose and methods of data collection will thus be important to the user of any GIS which draws on such information.

Census data

Rhind (1991) considers the different approaches which have been adopted for the collection of information about population size and characteristics. Although population registers are of significance in Europe, periodic censuses of population provide the most important source of socioeconomic information in many countries. The form and content of censuses varies widely, but they can result in extensive datasets, often collected and published for small geographic areas. The systems of zones created for the administration of censuses is frequently used for other data collection and reporting purposes. The TIGER system developed in the design of the 1990 US Census geography has already been introduced in Chapter 3. Here, the UK Census is used to illustrate the main issues in using census data as an input to GIS.

General introductions to the 1991 UK Census of Population and its geography may be found in Dale and Marsh (1993) and Martin (1993), but our primary concerns will be with the way the census data are referenced geographically and the means of extracting these data for GIS operations. Many other datasets are georeferenced by assigning observations into census zones, for instance in the calculation of mortality rates, expressed in relation to the census population, and in the allocation of postcode-based data to enumeration districts. Consequently, a very clear understanding of the census geography is required if such data are to be correctly interpreted.

73

The UK Census geography is a hierarchical system, as illustrated in Figure 5.2. The smallest data-collection zone is the enumeration district (ED), which represents the households visited by a single enumerator at the time of the Census. There were 130,000 contiguous EDs ('Output Areas' in Scotland) covering the whole of Britain in 1991. These EDs had an average population of around 400 persons and 200 households. Census-zone design represents a compromise between a number of conflicting criteria. It is desirable to avoid excessive variation in zonal area and population whilst maintaining reasonably compact shapes. If the reported data are to be of use, it should be possible to aggregate the smallest census zones neatly into larger administrative regions, and zone boundaries should follow obvious physical features as far as possible. The situation in different countries varies considerably: in France, there are no digital boundaries for the smallest units in the census-zone hierarchy, while in Portugal a set of digital boundaries has been created for the 1991 census similar to that used in the UK (Arnaud, 1993). In the US, the street segments which make up the TIGER database are the building blocks for census-zone boundaries, while in the UK, EDs tend to be more irregular, varying considerably in both shape and size. Under any such system, each zone may include large areas of open land, industrial and commercial properties and other non-residential land uses, giving a far-from-uniform within-zone spatial distribution of population. As the geography of the population changes between censuses (due to demolition or new residential development), so it is

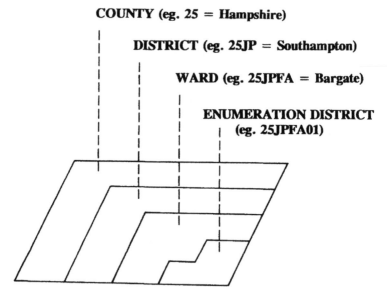

Figure 5.2 UK Census geography

74

necessary to change the census geography. A GIS would provide an ideal environment for the design of census geography, but, for the censuses of the early 1990s, census geographies have largely been designed manually, and the resulting boundaries and statistical data have provided an input-data source for GIS applications.

As illustrated in Figure 5.2, English and Welsh EDs may be aggregated to wards, wards to local-authority districts, and districts to counties. A hier-archic labelling system exists for 1991 zones, such that 25JPFA01 repre-sents an ED (01) in ward FA (Bargate), in district JP (Southampton), in county 25 (Hampshire). A similar scheme was used for the 1981 zones, and an additional character indicates those zones whose boundaries are unchanged between the two censuses. In 1981, the only geographically referenced digital data were a single population-weighted centroid location for each ED, and a set of digitized zone boundaries at the ward level. For 1991, there has been considerable expansion in the range of digital geo-graphic data available, largely reflecting the increased use of GIS. The ED centroid locations have been supplemented by two commercially produced sets of digital ED boundaries and a ward-boundary dataset created by the Ordnance Survey, together with a new directory of EDs and postcodes, which includes postcode locations and is discussed further below.

The centroid locations are included with the census Small Area Statistics (SAS), which is the statistical dataset containing census counts for each ED. This location is expressed as an Ordnance Survey grid reference (OSGR), to the nearest 10m in urban areas and 100m elsewhere, and is designed to represent the population-weighted 'centre of mass' of each ED. Centroid locations were determined by eye by the staff of the Office of Population Censuses and Surveys (OPCS) at the time the census boundaries were drawn up. It is very important to note that these are in no way geometric centroids, and that their locations were not determined by any directly reproducible means. Nevertheless, in an ED where the settled area com-prises only a relatively small proportion of the total land area, the centroid location should fall in that settled area. In uniformly settled zones, the centroid should be close to the centre, and in large zones containing a number of scattered settlements, the centroid should fall in the largest of these. The census centroids for zones at higher levels are spatial means, derived from the locations of the centroids of their constituent EDs.

The digital boundary data are available in a range of commonly used GIS data formats, and have been widely distributed. Some sets of boundaries have been sold packaged together with simple ED mapping software such as MAP91 which produces a range of shaded-area maps. Users of GIS software have the opportunity to recombine lower-level boundaries and aggregate data to new areal units, and to integrate the census data with other geographically referenced datasets.

Postcode-based data

Considerable and growing use is made of postal delivery code systems as a means of georeferencing socioeconomic information. Such codes have the important advantage that they are widely known, permitting relatively easy data collection, and yet relate to well-defined locations (essential for mail delivery purposes). In the US, the ZIP and 'ZIP Plus 4' codes, and in the UK the unit postcode, provide unique references to relatively small geographic areas such as one side of a street between two intersections. Most countries now have some form of (usually numeric) system for speeding the delivery of mail by providing reference codes for relatively small geographic areas. Thrall and Elshaw Thrall (1994) describe the US ZIP Plus 4 coding system and the various ways which are now available for the assignment of grid references to addresses using these codes. This increased use of postcodes was indeed one of the recommendations of the Chorley Committee in the UK (DoE, 1987). A comprehensive explanation of the history and operation of the UK postcode system may be found in Raper *et al.* (1992). There are around 1.7 million postcodes in the UK, covering approximately 25 million mail delivery addresses. Of these, 170,000 are 'large-user' addresses, receiving in excess of 25 mail items per day, but the remainder are mostly residential addresses. Each postcode is an alphanumeric code consisting of postal area and district (outward) codes, and sector and unit (inward) codes, as illustrated in Figure 5.3. The full unit postcode represents the smallest indivisible part (e.g. a single terrace of houses) of a delivery postman's 'walk', and contains on average fifteen addresses. In England and Wales there are no definitive boundaries for unit postcode areas, which are defined solely by a list of street addresses, although in Scotland digital postcode boundaries have been created. This absence of boundaries was one of the main obstacles to the creation of a census geography to match the postcode geography. Digital boundaries and paper maps are commercially available for most zones in the postcode hierarchy, and new data products are appearing with increasing levels of geographic detail.

In 1976, postcodes were used by the Department of Transport as the georeferences of trip ends in a nationwide transportation survey (Sacker, 1987), and a computer file giving a 100m Ordnance Survey grid reference (OSGR) for each unit postcode was assembled. Officially, the OSGR in the Central Postcode Directory (CPD) represents the lower left corner of the 100m grid square containing the first address (alphanumerically) in the postcode. This file formed the basis for the present CPD, which includes a ward code with each record, and is updated every six months as demolition or new development require changes in the mail delivery system. The Post Office also maintains a postcode address file (PAF), which contains the officially designated postal address and postcode of every address in the

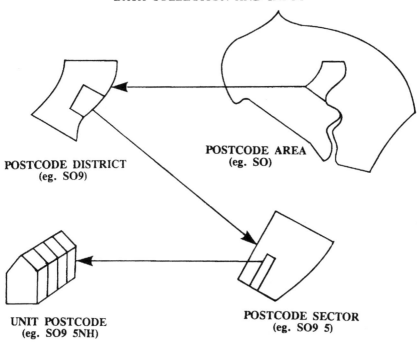

POSTCODE DISTRICT
(eg. SO9)

POSTCODE AREA
(eg. SO)

UNIT POSTCODE
(eg. SO9 5NH)

POSTCODE SECTOR
(eg. SO9 5)

Figure 5.3 UK postal geography

UK (Post Office, 1985). This allows any street address to be allocated the appropriate postcode and thus linked to other records via the CPD. The main priority of the Post Office is the delivery of mail, not the maintenance of these files, but as increasing use has been made of the postcode system for non-mail uses, attention has been given to improving the accuracy of the OSGRs in the CPD, and corrections are made at each update. The results of an evaluation of the 1986/1 version of the CPD showed that 72 per cent of the grid references were accurate to within 100m, 93 per cent within 400m and 95 per cent within 900m of their true location (Wilson and Elliott, 1987). Since that time, extensive revisions and corrections have significantly improved the locational accuracy of the CPD, but the true level of error remains hard to assess (Raper *et al.*, 1992). Due to the use of the nearest 100m grid reference to the south-west of a postcode location, spatial accuracy may be increased for some applications by adding 50m to both x and y coordinates, to reduce the average spatial error of the CPD references.

For the 1991 Census, it was proposed that EDs should be aggregates of unit postcodes (see below), as was the case in Scotland in 1981. This would have greatly enhanced the facility to allocate postcode-referenced data to the census. However, due to the time and cost involved in defining unit postcode boundaries, this did not take place and there is no correspondence between census and postcode geographies. Instead, a new directory of EDs and postcodes was created which contains a record for each unique intersection of an ED and a postcode. These small overlapping units have been termed part postcode units (PPUs). The structure of the directory is illustrated in Figure 5.4, and the relationship between postcode and census geographies is further discussed in Martin (1992). For each PPU, the identity of the ED and postcode are given, together with the identity of the ED to which the PPU would belong if whole postcodes were assigned to the EDs in which the majority of their population fell (termed the 'pseudo-ED', or PED). The grid reference of the postcode from the CPD is given, together with the count of the number of households in each PPU. A final item of information is the imputation code, which indicates the number of census forms returned without a postcode, from which the postcode was imputed from their neighbours. Although less than ideal, this directory does provide a more direct method for matching postcoded data to the census than was previously available, and permits users to more precisely estimate the characteristics of areas based on postal geography. Plate 3 illustrates the ED boundaries and 100m postcode locations for an area of central Cardiff. The regular grid structure imposed on the postcode locations by 100m referencing are clearly apparent. As Barr (1993b) notes, many of the processes involved in the matching of postcoded records to census geography are list-matching operations which may be performed by non-geographic database software and do not actually require the spatial processing capabilities of a GIS.

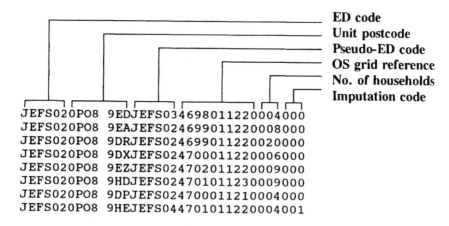

Figure 5.4 Structure of the UK directory of enumeration and districts and postcodes

DATA COLLECTION AND INPUT

Address- and individual-level databases

An important development in the late 1980s was the development of a new product by Pinpoint Analysis Ltd, known as the Pinpoint Address Code (PAC). This product contained a single grid reference for each property, linked to the Postcode Address File. The advantages of such a high-resolution dataset were clearly demonstrated by Gatrell (1989), who compared the performance of the CPD and PAC data in distance-based and point-in-polygon assignments of properties to census EDs and wards. Unfortunately, the PAC data were expensive and failed to achieve national coverage. In addition, any dataset of this size suffers from the need for constant revision in the light of changing postal geography, construction of new properties and demolition of old ones. For various reasons the PAC failed to become the national property-level referencing system which was so clearly needed. In April 1993, Ordnance Survey launched its new ADDRESS-POINT product, starting in selected urban areas, and with the objective of achieving national coverage by the end of 1995. ADDRESS-POINT provides a national grid coordinate to 0.1m resolution and a unique reference (the 'Ordnance Survey ADDRESS-POINT Reference', or OSAPR) for each postal address (Ordnance Survey, 1993). The product has been developed on the basis of the Post Office's PAF, which provides the definitive listing of the 25 million postal addresses in the country. ADDRESS-POINT is undoubtedly one of the most significant geographic datasets ever to be released in the UK, and provides a definitive basis for the georeferencing of socioeconomic data. Although the data are currently expensive, the existence of a single dataset which provides a precise georeference for every address in the country is likely to have major impacts. There is currently no explicit link between the ADDRESS-POINT data and the census, but this problem may be overcome by further derived data products, as the definitive address locations may be easily allocated into any set of zones for which digital boundaries are available. Plate 4 illustrates the ADDRESS-POINT locations together with street centrelines and ED boundaries for the same region as Plate 3. The advantages of address-level georeferencing are made clear by the contrast between the two illustrations.

In addition to the applications already noted, many organizations hold personal- or property-level databases which are not explicitly georeferenced, or alternatively have detailed digital data which are not linked to any address database. Health authorities, local government and retailers, for example, hold extensive records which contain addresses for individuals, but which do not include grid references and are thus not directly amenable to mapping or geographic manipulation. Other organizations hold information which primarily concerns properties or land parcels, such as the rates registers maintained by local government and water companies, and the new council tax register in the UK. Such registers contain some form of unique property reference number (UPRN), which identifies each address or subdivision of an address, such as flats or bedsits in a large subdivided house. UPRNs may allow an organization to organize its billing or mailing operations and in many countries are used for cadastral purposes, but they do not contain any explicitly spatial component such as a grid reference. Postcoded records are helpful in any of these databases, but are still limited by the need for one of the georeferencing methods already considered. Additional problems with such registers concern definitions and access: for example, the council tax register contains only domestic properties, and is not available to the public in computer-readable form. Due to the specific purpose for which it was created, it uses different definitions of properties from those found in the earlier rates register, particularly regarding subdivided properties. An example of the structure of a UK rates register (containing fictitious addresses) is shown in Figure 5.5. The UPRN (first column) is a hierarchical number in which the first three digits (in this case, all 101) represent a neighbourhood within a city, the next four digits represent a specific street, and the next four a unique street address. The final two digits of the UPRN identify subdivisions within the same address, such as flats on different floors. It should be noted that, as in the figure, such registers frequently contain a variety of address formats, and must accommodate many different types of property, making automated analysis of the list very complex. There is no grid reference or other explicit locational reference in such a register. Martin *et al.* (1994) consider the many different digital databases which provide geographically referenced data for the inner area of Cardiff, and highlight the problems created by inconsistent definitions, incompatible UPRNs and restricted access to data. It is apparent from this work that the real obstacles to data collection and input are frequently organizational in nature, and concern accessibility and cost. The integration of these types of information also raises important issues concerning data standards and the protection of data about individuals, both of which are dealt with more fully in later chapters.

Many specialized databases come into existence either as the result of specific surveys or from the maintenance of customer records by organizations interested in the characteristics of their clients, and thus provide important socioeconomic information about named (or traceable) individuals. The latter type would include hospital patient administration systems (PAS), local-authority electoral registers and community charge registers (Burton, 1989), and store-creditcard holder records. These data are frequently address-referenced, and thus may potentially be aggregated to areal units such as store catchment areas or census zones via the CPD. Often, these sources contain information which does not exist in generally available datasets. Some, such as health records (particularly mortality and morbidity registers), have been collected over a period of years mainly for

Reference	Type	Address	Location
101012000002410	67 HOUSE	24 JUTE STREET	CARDIFF
101020000002120	161 MAISONETTE	28 STATION STREET	CARDIFF
101020000003910	161 MAISONETTE	FIRST & SECOND FLOORS	39/40 STATION STREET
101030000004010	480 MAISONETTE	ST JOHNS CLERGY HOUSE	REAR 40 EDWARD STREET
101030000004301	157 MAISONETTE(GF & 1F)	43 EDWARD STREET	CARDIFF
101030000004803	108 FLAT	THIRD FLOOR	48 EDWARD STREET
101030000007000	4 CATHEDRAL	ST DAVIDS CATHEDRAL	EDWARD STREET
101036000002402	301 FLAT	SECOND FLOOR	24 CHURCH WAY
101068000001509	47 FLAT	NINTH FLOOR	BEECH HOUSE
101074000000100	72 HOUSE	PUBLIC BATHS	ROMFORD CRESCENT
101090000000303	69 FLAT(THIRD FLOOR)	3/4 LOW STREET	CARDIFF
101090000002604	59 FLAT(FOURTH FLOOR)	26 LOW STREET	CARDIFF
101090000002805	50 FLAT(FIFTH FLOOR)	28 LOW STREET	CARDIFF

Figure 5.5 Example of a property register: a UK rates register

reporting and the compilation of statistics. There is much potential for geographic analysis of these data within GIS. Others, including income, savings and spending information, may be of considerable operational or commercial interest to a wide variety of organizations. GIS techniques offer the potential for more accurate health-care planning, direct mailing, or academic research, and thus these databases are extremely valuable (Brown, 1991). The existence of such information also raises serious issues of confidentiality, which are addressed more fully in Chapter 8.

Socioeconomic data from remote sensing

On pp. 21–25, an overview of remote sensing (RS) was given, in the discussion of the development of image processing. These systems are an increasingly important source of data for GIS. A notable application of RS data has been in the monitoring of urban areas and in population estimation (for example, Duggin et al., 1988; Lo, 1989; and Sadler et al., 1991). A consistent problem of image classification is the absence of any 'general knowledge' within the system as to which areas might reasonably be classified as urban. The need for meaningful ancillary data for correct interpretation has been reflected in recent attempts to find appropriate surface variables for inclusion in the classification exercise (Mason et al., 1988; Langford and Unwin, 1994). Despite these difficulties RS data have considerable utility in the mapping of land cover, and their timeliness makes them a potential basis for the updating of other data sources where ground surveys are carried out infrequently. The use of RS data as an information source for land cover mapping within GIS is explained in some detail by Avery and Berlin (1992).

One of the main problems with the pixel-by-pixel approach to the classification of urban areas is that the texture of land uses is typically diverse, even at the level of the individual pixel. A consequence of this is that the spectral signal of individual pixels may represent the presence of a mixture of different objects on the ground (buildings, gardens, roads, etc.) and not be clearly assigned to any one land-cover class. This is a common difficulty with urban RS analyses and is especially prevalent in suburban areas and lower-density settlements. In many countries, detailed and routinely collected data such as censuses are able to provide high-quality geographic information about population distributions. However, remotely sensed data may have an important role in population estimation where routine data collection is inadequate. Lo (1986) considers the various ways in which satellite imagery and aerial photography may be used in such applications, including the counting of individual dwelling units, and population inference from the measurement of different land-cover areas. Limited socioeconomic information may also be obtained by inference from environmental quality and residential density. Interestingly, remotely sensed data are also a powerful archaeological source in respect of historical land-use and settlement patterns.

The data-collection transformation

Before moving on to consider the technology available for the input of these sources of information to GIS, note should be taken of the trans-formations which have taken place in the collection of the data mentioned above, changing both its fundamental spatial and attribute characteristics.

In most of the above cases, information for a number of discrete individuals is aggregated to areal units. In the case of census data, this aggregation is explicit; in the case of address-referenced information, a number of individuals may be expected to live at the same address and thus share the same postcode. The areal units to which the data are aggregated may in turn be geographically referenced in a number of different ways: either defined by boundaries, which will enclose all individual cases (census zones), by a point which is a reference to a defined area (census centroids), or by a point which is a reference to a set of locations which do not fall within a defined boundary (postcodes). The newer address-level datasets approach the provision of a unique grid reference for each household, but it must be remembered that there may be many diverse households at the same address, and many diverse individuals within any household. Each of these data collection scenarios will have different implications for the use of the data in spatial manipulation operations such as overlay with some physical or administrative region. In the case of the remotely sensed data, individual records are never collected, and the data will only ever have meaning at the level of the ground resolution of the satellite, which may be further obscured by the nature of the method used for image classification.

None of these data-collection methods, except perhaps the recording of locations of individual households, is without some form of object-class transformation. In most cases this is aggregation from point to area classes. All are subject to a degree of spatial error (precision of grid references, accuracy of directories). The recording of essentially continuous real-world phenomena such as boundary lines as series of points, either surveyed in the field or drawn on a base map, involves generalization which affects the lengths of routes and the areas and perimeters of zones. Attribute informa-tion is generally held in relation to each spatial object (e.g. population counts for each zone), and is susceptible to a variety of well-known data-collection errors. Additionally, geometric errors in the recording of zone boundaries may make attribute values incompatible with the new (incor-rect) areal units.

DATA INPUT

Data input in vector and raster forms are here considered separately, as this is a helpful distinction in terms of the techniques used for data entry. However, there is no general distinction between the two data types in terms of the nature of the transformations undergone by the data. Particularly in the context of vector data, the input processes may generate major logical and locational errors, which require special techniques for data verification. Often this involves the use of software tools which are more commonly associated with data-manipulation operations, but, as they are brought into operation before the input data become part of the digital database, it is appropriate to consider them in the context of data input. The input of attribute data is again considered separately, as many sources of attribute information exist which may be appended to an existing spatial database.

Vector digitizing

The input of vector data most commonly consists of the recording in digital form of the coordinate and topological information from a source diagram, usually a paper map. The availability of relatively low-cost global positioning systems (GPS) and total surveying stations is making the resurveying of topological information with direct digital data input a practical option for many organizations. Nevertheless, most national topographic databases and existing records must be converted from paper to digital form. This process of digitizing may be performed by automated means, but is more commonly accomplished manually and represents the major bottleneck in the creation of digital map databases (Chrisman, 1987). Manual digitizing involves the operator tracing the source document with a cursor on the surface of a digitizing tablet (as illustrated in Figure 5.6). This is an electronic or electromagnetic device which is able to detect and transmit the position of the cursor with a very high degree of precision. Once the source document is positioned on the surface of the tablet a number of control points are recorded, to which real-world coordinates may be given. This operation is generally referred to as 'map registration' (although all the techniques we shall be discussing tend to have slightly different names according to the software being used). By defining the locations of a number of control points in terms of some real-world coordinate system, all recorded points may be transformed into that system from the tablet's internal system (typically in millimetres or inches). Data are passed by the tablet to a host computer, where the data are stored on disk prior to verification and entry to the database. Some geometric reprojection (e.g. affine transformation) will also be necessary at this stage to overcome distortion in the source document and to correct for

84

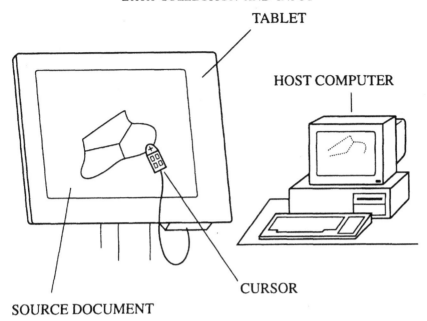

TABLET

HOST COMPUTER

CURSOR

SOURCE DOCUMENT

Figure 5.6 Manual digitizing

poor positioning on the tablet. Coordinates may be recorded in either point mode (the operator being required to press a cursor button for each coordinate) or stream mode (in which coordinates are recorded continuously). In most applications, the more laborious method of point-mode digitizing is employed, due to the massive volumes of data accumulated in stream mode.

The quality of the source document is a major constraint on the digitizing task, as torn or stretched sheets will cause errors in the digital database, especially if a number of neighbouring sheets are to be matched together. Poor linework and labelling will lead to increased operator errors in what is already a time-consuming, tedious and error-prone process. Many of these issues are illustrated by Fisher (1991), who notes the many ways in which conventional paper maps, designed for other purposes, may be poorly suited as data sources for GIS. Maffini *et al.* (1989) present the results of a study of digitizing error in which they identify a number of specific contributory factors such as line complexity, source-document scale and time constraints. From this study they conclude that a degree of inaccuracy is inevitable, and that the only realistic 'solution' is to develop methods to understand and allow for this error in GIS software. Despite the acknowledged dangers, distortion-free source documents are not often available, and many applications rely on the use of photocopies or hand-drawn detail on general-purpose map sheets. This is especially the case with many socioeconomic boundary maps, which tend to be defined by reference to existing paper topographic maps (e.g. OPCS, 1986).

Control of the digitizing process is typically via a menu area on the surface of the tablet or by the use of a multi-button cursor. Areas of the menu or buttons on the cursor are assigned to particular commands (e.g. 'erase last point', 'start new line', 'exit', etc.), giving the operator access to the digitizing system's commands without having to move from the tablet to the keyboard. Modern digitizing software displays the data currently being digitized on the computer screen to assist the operator. Although the large-scale use of manual digitizing is far from satisfactory, few significant advances have been made towards the widespread automation of the process. Large mapping organizations often employ shifts of manual digitizer operators working round the clock, and commercial agencies exist which offer contract digitizing of clients' source documents. The separation of spatial and attribute data structures means that input of these data may be performed independently. Linkage usually relies on the encoding along with the spatial information of a 'key attribute', which may be used to search for that object in associated attribute tables, or in pre-existing databases, as mentioned above. Digitizing is essentially a scale-dependent process. The digital representation can never contain greater detail or achieve higher locational accuracy than the original document, and the degree of line generalization which takes place during input is under the subjective control of the operator. Digital maps should not be used or reproduced at a scale larger than that of the diagram from which they were created.

Automated input of vector data may be achieved either by the use of raster-based scanners and subsequent conversion of the data to vector form, or with vector line-following devices. Both require the existence of very high-quality source documents and a high degree of operator intervention in line labelling and scanner supervision. Both approaches are best suited to relatively simple cartographic data such as contour lines, which may be clearly distinguished from the map background. The considerable expense of such specialized hardware and software has limited their use to commercial digitizing agencies and to organizations with large data-input requirements of a suitable type, such as the OS (Fraser, 1984). Significant cost and size reductions have been achieved in raster picture-scanning technology, but the algorithms for the conversion of such images to cartographic-quality vector data remain problematic.

The discussion in the following chapter will illustrate the degree to which the input data must be carefully structured in order to provide a useful digital model of the geographic world. Although there are a wide variety of options for the structuring of vector data (e.g. segment-based, polygon-based), all digitizing software requires functions for the verification and structuring of the data in some form. Some systems require the operator to give labels to map features at the time of digitizing, others accept a mass of unlabelled 'spaghetti' which must subsequently be verified and labelled. Using either approach, considerable care is required to avoid errors such as mislabelling, digitizing features twice and missing features altogether, and verification software seeks to identify and resolve these problems. The degree of sophistication achieved varies considerably between systems.

Point-referenced data may be created by field surveys, but in the context of socioeconomic applications are frequently input directly from existing files, such as the zone centroids for census zones and postcode locations from the CPD. These may be held as point-based layers within the GIS, or used as a basis for the generation of new vector features by manipulation. Examples of these techniques include triangulation and polygon generation around points: operations which are explained in Chapter 7.

Verification

In addition to inaccuracies in the source document, a number of common data errors arise from the digitizing process itself. It is necessary to correct these (preferably automatically) before the digitized data are entered in the spatial database. The main error types are illustrated in Figure 5.7. The basic causes and software solutions to these situations are considered here with reference to the error numbers in the figure. Segments here refer to the series of lines defining the boundary between two adjacent polygons. Nodes are the points which define the ends of segments.

It is extremely unlikely that the operator will choose precisely the same location on the digitizing tablet when recording the same point at different times as part of adjacent polygons. Due to the high precision of the tablet, the same real-world point may thus be entered as more than one database entity. Where closed polygons are being digitized, this will lead to the creation of sliver polygons (error 1). In line-segment digitizing (errors 2, 3), only the nodes need be digitized more than once, and the identification of all coordinates representing the same point and their correction to a single database coordinate pair is referred to as 'node snapping' (also 'node capture' or 'feature tie'). If sufficient processor power is available, it may be possible to test all previously digitized nodes during the digitizing process to see whether they fall within a certain radius (tolerance) of the node last digitized. If there are other nodes within the tolerance distance, these may all be snapped to the same location, if required. More commonly, verification operations such as node capture are performed as a separate stage after digitizing.

Once nodes have been snapped and segment labels added, polygon construction is performed to test for completeness in the database. All segments associated with each polygon are assembled in order to check that polygon boundaries are complete. Errors 4–6 may be due to missing

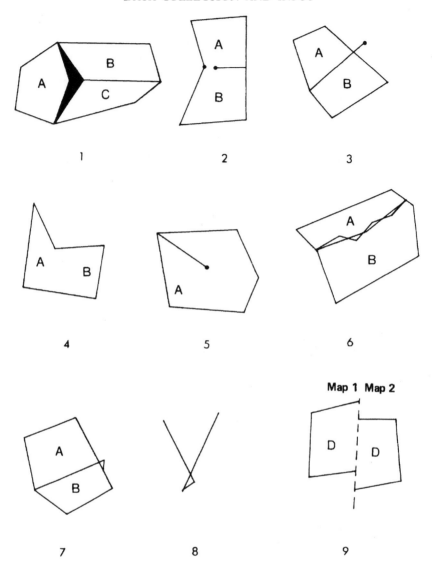

Figure 5.7 Common vector data errors

segments, wrongly labelled segments or segments that have been digitized twice. Often, it will only be possible to resolve these problems by reference to the source document and editing the database. Good verification software will offer comprehensive facilities for the manual editing of digitized data. Errors 7 and 8 ('weird polygons') may occur due to careless digitizing, often caused by poor-quality source documents.

The final error indicated (9) relates to the creation of multi-sheet databases and is only one of a range of related errors. If the digital map base is to cover an extensive area, it will frequently be derived from a number of separate paper maps, and the data from each of these sheets must be matched together. This involves snapping edge points, as with nodes (above), and checking for consistency in labelling to allow the reconstruction of polygons falling across the sheet boundaries. Multi-sheet software systems adopt different strategies for the storage of this type of data, but usually store edge-matched sheets in separate indexed files, to allow the construction of output maps which fall across boundaries without holding extremely large spatial data files. Edge-matching problems may be particularly severe where significant linear features such as contours or roads fall across the map sheets, as there will be no simple solution to mismatches without obvious distortion or discontinuities in the data. Polygons may be incorrectly labelled by automated attempts to resolve boundary mismatches at map edges.

Attribute-data input

Once a vector database is established, attribute data may be imported from a wide variety of sources which relate to the spatial objects in the digital map. Using the concept of a key attribute, introduced above, it is a relatively simple matter to add extra fields of attribute information as they become available, usually as output from other information systems. Many organizations will have their own attribute data maintained within a proprietary database management system (DBMS), and links will be established with the GIS, either by the transfer of selected data to its internal database or by sharing data dynamically between the two systems. In addition, examples are given here of sources containing information of interest to socioeconomic GIS applications. These data may be associated with either vector or raster coverages, but the entities which are found in the attribute databases are more likely to directly correspond with vector objects in the GIS, whereas their association with the cells of a raster coverage will require more complex manipulation.

In the case of 1991 UK Census data, around 9,000 census counts are available for each census enumeration district (ED) in a dataset known as the Small Area Statistics (SAS), and a more comprehensive set of around 20,000 counts is available at the ward level and above, being known as the

Local Base Statistics (LBS). A range of other specialist products concerning migration, travel to work, an anonymized sample of individual records, and a longitudinal dataset are also available, but these have either been less widely used or are less suited to analysis within GIS. These datasets are provided by the census office, OPCS: and two widely used software packages, SASPAC91 and C91, are available for the manipulation of the SAS and LBS. These data-retrieval packages allow the creation of new variables by simple calculation and the creation of new zones by aggregation of existing ones. In addition, they provide some tools for matching census- and postcode-referenced datasets. Although these systems are used widely for census data reporting, particularly in the public sector, more complex geographic analysis usually requires the importing of the census data into GIS, as attribute information corresponding to one of the digital boundary datasets mentioned above. As Barr (1993b) stresses, simple mapping and reporting of census data does not require a GIS, and there has been evidence of the use of complex GIS for handling census data where much simpler software would have sufficed. Nevertheless, the potential for the development of innovative forms of spatial analysis of population data using GIS should not be underestimated (Martin, 1995). It is also important to recognize that attribute data of this kind may bring many non-locational errors into the GIS database. Census counts are subject to modification in order to avoid inadvertent disclosure of information about individuals, and data for some small areas are restricted. Censuses are subject to underenumeration, usually of specific sub-populations such as the homeless, and may include elements of data imputation (OPCS, 1992a). The UK Census shares with that in the US the feature that some counts are only available for a sample of the population, and thus sampling errors will also be associated with this part of the census information. Perhaps the most general problem with all counts derived from census data is that they are collected only infrequently, and data are therefore liable to be several years out of date before they are even available for processing within a GIS (Fisher, 1991).

The UK census data are publicly available, according to a scale of charges, which is calculated to recover some of the costs of data collection, a situation which also applies in Canada, for example. In the US, however, where census counts are available in computer-readable form for all 7.5 million census blocks, charges are simply intended to cover media costs, due to the different way in which public data collection is viewed (Maffini, 1990). For the first time in the 1990s, data from all these national censuses have been widely processed using personal computers, again indicating the massive decreases in cost of processing power which have continued to take place over the last decade.

Maffini (1990) and Rhind (1992) both stress the importance of government datasets in the development of GIS applications. Many of the data

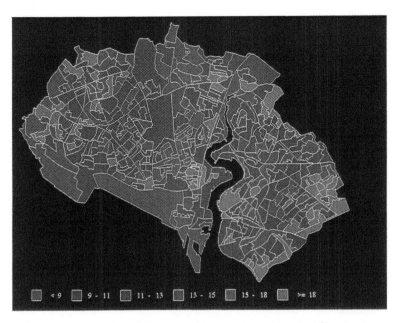

Plate 1 Percentage of population under 10 years of age in the City of Southampton (1991 Census)

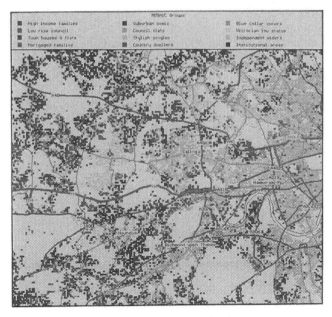

Plate 2 MOSAIC lifestyle groups in West London: a contemporary neighbourhood classification scheme. Copyright © CCN Marketing 1992, © Post Office 1992 and © Automobile Association 1992

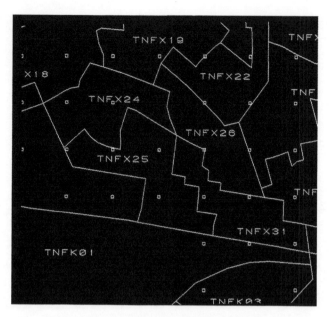

Plate 3 1991 enumeration-district boundaries and 100m postcode locations for Cardiff study area

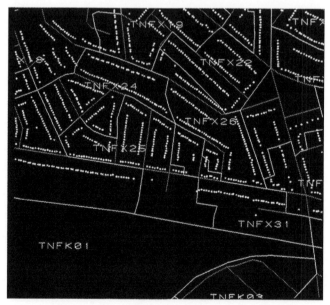

Plate 4 1991 enumeration-district boundaries, street centrelines and ADDRESS-POINT locations for Cardiff study area. ADDRESS-POINT data used by kind permission of Ordnance Survey © Crown Copyright

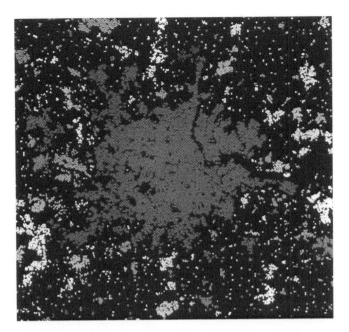

Plate 5 Population density surface model of Greater London
(1991 Census)

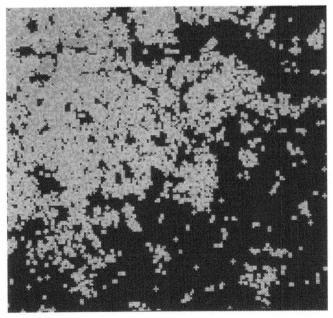

Plate 6 Population density surface model of Greater London (inset)
(1991 Census)

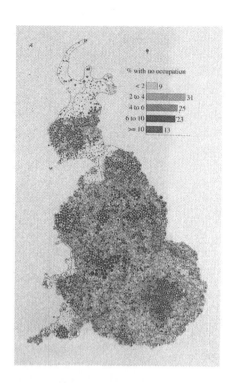

Plate 7 Cartogram representation of ward
populations with no occupation
(1991 Census)

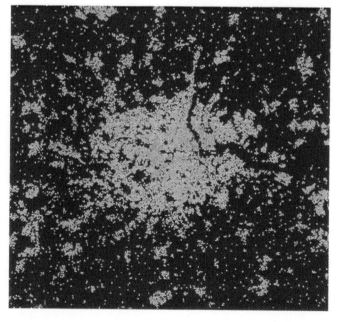

Plate 8 Discrete populated settlements in the Greater London region

series routinely collected by national governments have important geographic dimensions and are frequently available by quite small geographic areas. Government policies towards the ownership and dissemination of such data have thus proved very influential on the nature of GIS applications which have been possible. In the UK, an important source of geographically referenced official statistics is the National Online Manpower Information System (NOMIS), described by Blakemore (1991). The system has been developed at the University of Durham and has been in full operational use since 1982. Data are available online to registered users via telephone connections or academic and commercial computer networks. There are five major NOMIS databases: demographics, projections, employment, unemployment and vacancies. Most of these datasets are available at a range of geographic scales, with wards as the basic 'building block' of the system. Aggregations to user-defined regions are also possible. Despite the ability to aggregate data from basic areas, this type of system serves to illustrate the problems of obtaining population-related data which relate to incompatible areal units. This is a key obstacle to the creation of area-based models and requires very careful selection of the base areas. In all countries, government statistical data seem likely to remain as one of the key sources of socioeconomic information; their value is greatly increased by integration with organizations' own operational information within GIS. At a European level, the Statistical Office of the European Union (EUROSTAT) is currently working towards a European statistical system in which standardized geographically referenced data are collected and integrated for the EU countries. In the future, such international initiatives may provide important data inputs for new GIS applications. In addition to these public-sector organizations, there has been steady growth in the commercial supply of general-purpose socioeconomic information such as lifestyle indicators and the results of consumer surveys, with an emphasis on value-added products where some additional integration or georeferencing has been performed.

Some of the datasets already mentioned in the context of locational referencing are also important sources of attribute information. Cadastral and rates register databases provide useful clues regarding income levels and housing quality, which may be associated with existing property records within a GIS. Most GIS-using organizations will have important data series of their own which may provide additional fields of attribute information. Data from systems such as SASPAC91 and NOMIS must be extracted in an appropriate file format, transferred to the machine running the GIS, and imported in some form to the system's database. Increasingly, GIS installations may be expected to share databases with other information systems, as the costs of downloading and duplicating data within GIS are in many cases prohibitive. Distributed databases will become more common, as more processing power is available at the users' own workstations and there is a need for centralized data holdings, shared by many users with their own specialized software.

One final type of data which is worthy of mention under this heading is that of 'metadata'. Metadata may best be described as 'data about data'. As the quantity of digital spatial data has increased, there has been increasing demand for knowledge about what these data contain, the rules and definitions used in their compilation, explanations of missing values, and data quality and coverage. These are the types of information which comprise metadata (Laurini and Thompson, 1992) and which may be incorporated as attributes within spatial datasets. As has already been stressed here, the nature of the data-collection and input operations will have significant impacts on the ways in which it is appropriate to use a particular digital data product, but these details have not usually formed part of the data themselves, and are thus easily lost when the data are passed to another user or transferred across a computer network. Metadata may take the form of information which is included with every digital dataset, describing its creation and update history, but the concept also encompasses digital data catalogues and indexes, which describe what is available, together with important details concerning creation and usage. Such information may be updated as data undergo further manipulation and reprocessing. Most existing GIS software contains few, if any, tools for the handling of such metadata, and even where manipulation operations on each dataset are logged this information is not always directly available to the user. The need for adequate ways of providing metadata (or 'meta-information') is increasingly being seen as a crucial component of the development of data standards, discussed in Chapter 8. Burrough (1994) identifies the difficulties of presenting information on accuracy and uncertainty in this way. Not least of these is the fact that the inclusion of even small quantities of metadata, such as a standard deviation for each quantitative attribute, would double the size of the database.

Raster data input

Raster data are generally obtained as output from another information system or spatial modelling program. Remotely sensed images may be imported directly from image-processing systems and stored as raster coverages within the GIS. Most GIS software does not incorporate the sophisticated functions required for image rectification and interpretation, so the data imported are frequently in the form of classified images, with all the input operations being performed in the image-processing system. However, the growth of 'integrated' systems with both IP and GIS functionality is serving to rapidly reduce this gap in functionality. The data-storage structures adopted in GIS databases involve more sophisticated structuring of the data than those commonly encountered in IP. RS images are typically held as full matrices of single-byte codes, giving possible values in the range 0-255 for each cell in the image. GIS must provide the facility to hold values outside this range and adopt various strategies for reduction of redundancy and increase of access speed in the database, as discussed more fully in Chapter 6.

One recent growth area has been the import of scanned raster map images in order to provide background information for utility-management applications. Such images are created from existing paper map products and provide a visual backdrop without associated attribute information. These now form a major data product for organizations such as the UK Ordnance Survey (Rhind, 1993). Raster data are attractive for this kind of application because they can be obtained automatically and without the need for constant operator intervention, thus providing extensive geographic coverage relatively cheaply. The scanning of map images for the production of raster data coverages is less problematic than for vector applications because there is no need to reconstruct topological relationships within the scanned data. Jackson and Woodsford (1991) consider scanning devices, and note that all scanning involves systematic sampling of the source document, by transmitted or reflected light. The resolution of the scanning device is thus crucial to the quality of the output data, and higher resolutions lead to rapidly increased data volumes. Higher resolutions will generally be required for data that are to be vectorized than for raster applications such as background mapping.

Many raster GIS coverages are constructed from vector input information by rasterizing line and area data or interpolation of point data sources. The techniques for these types of operation are discussed in Chapter 7. No separate attribute information is required, as the attribute values are integral to the raster data, and input is generally on a sheet-by-sheet basis, unlike the individual entity encoding of vector input.

The data-input transformation

In most of these data-input operations, the most important transformation processes at work on the data are geometric. Data input consists basically of the encoding in digital form of information already collected. Object-class transformations would not normally be encountered at this stage. Attribute values imported directly from other information systems will not be directly affected, but they may no longer be entirely accurate if the features to which they relate have been altered. The implications of this kind of error are complex. If a population count is obtained for a census zone A, but that zone has been incorrectly digitized, the system will still return a correct answer to a question such as 'what is the population of zone A?'. However, if a question is posed according to spatial criteria, such as 'what is the population of new zone B?', where B is a newly created spatial feature including the region digitized as zone A, the answer to the question may be incorrect, and there is no way of determining the true answer without reference to information held outside the system.

SUMMARY

In this chapter we have considered the two important data-transformation stages of collection and input to GIS. In socioeconomic applications, GIS usually rely on data already collected for some other purpose, often acquired in the operation of an organization which keeps records of its clients. Input of these data in many cases simply involves the importing of records from databases existing within other systems. However, an implication of this is that the spatial model within the GIS is heavily dependent on the data collection methods used, in which accurate digital representation of the data is rarely a consideration. Major difficulties surround data for incompatible areal units and data for which there is no direct form of georeferencing. In these cases, various types of error and uncertainty may be introduced by the process of assigning spatial coordinates to the data. Sometimes, aggregation occurs which makes impossible the identification of individual entities in the final model, and in these cases all subsequent analyses are restricted to the level of the base areal unit. Where manual vector digitizing is involved, the digitizing process is itself highly error-prone, and complex software is required to trap and resolve these errors. In the following chapter, we shall examine the structures used for the storage of these data once input and consider their characteristics as models of geographic reality.

6

DATA STORAGE

OVERVIEW

One concept, which has been stressed throughout our discussion, is that of the digital map base as a special kind of 'model' of geographic reality. We have seen how data from a variety of sources may be input to the computer, and it is now necessary to consider the structures which may be used for the storage of these data. Any such database can be viewed as a computerized record-keeping system, and just as a badly organized records office will tend to sprawl and take a long time to answer a simple enquiry, so the same principles apply here. The framework chosen will have important implications for ease of access to, and size of the data, and different structures exist which will be best suited to different types of application.

Desirable properties of any data structure are that it should be compact, avoiding duplication and redundancy as far as possible, and that it should support easy retrieval of data according to whatever criteria the user may specify. The basic building blocks of such systems are entities – the objects about which the data have been collected – and the relationships between them. Geographic data are unique in the sense that every feature will have locational relationships (e.g. bearing, distance, connectivity) with every other feature, in addition to the logical and functional relationships (e.g. employer–employee, customer–purchases) which may be expected to exist in non-geographic databases. This view of the world in terms of features and their attributes is essentially a vector-oriented view, and is reflected in the structure of many vector-based GIS. These have a specially structured spatial database, in which the locational characteristics of features are stored, and a separate attribute database for non-locational characteristics. In some GIS packages, this distinction is explicit, and the storage structure used for attribute information may be a commercial database management system (DBMS) in its own right. Different approaches to vector data storage are explained in the following section, and then a basic introduction to attribute database handling is given. Those wishing to learn

95

more about multi-purpose DBMS packages should consult a text on general database management such as Date (1990).

Some consideration is also given here to object-oriented approaches to modelling geographic entities. In this increasingly popular technique for spatial data structuring, all geographic entities are classified into a range of object types (e.g. 'house', 'field', 'street'), to which a set of properties and operations can be applied. These objects are the basic components of the data model, and the structure can be applied to vector, raster and attribute data types.

The raster approach to data representation and storage imposes rather a different view of the world, as it is basically coverage- and not feature-oriented. Every cell in a coverage has some attribute value, and individual features are not separately recorded. A land parcel may actually be represented by a group of adjacent cells whose attribute values relate to that parcel, but there is no sense in which the parcel itself is a database entity. Vector and raster representations of the same spatial object are illustrated in Figure 5.1. Consequently, raster storage is much simpler than vector storage in terms of the organizational structures required, a database merely consisting of a group of georeferenced coverages, each of which represents the values of a different attribute at every cell location.

Finally, mention is made of the triangulated irregular network (TIN), a method for storing models of surface-type phenomena. Although commonly applied to models of the physical landscape, the technique is of interest due to its variable resolution, providing more information in regions of greater surface variability.

A further distinction necessary here relates to the need for efficient and compact data storage, mentioned above. In addition to the way in which the contents of the database relate to the geographic world, this may be achieved in the techniques by which they are encoded in the computer. This applies to all the above methods for structuring geographic data. Raster representations in particular, although conceptually simple, potentially include much redundant data, and considerable effort has been directed towards finding the most efficient ways of encoding these data to reduce memory and disk space requirements.

VECTOR DATA STORAGE

Maffini (1987) notes that vector methods may impose subjective and inexact structure on the landscape, but are more suited to situations where there is a need for precise coordinate storage. Important topological information may also be encoded which is very hard to record using raster data structures. The recording of socioeconomic phenomena has generally employed vector techniques, due to the precise nature of the boundaries used, for example, for census-area definition. A difficulty arises because the precise encoding relates only to the boundaries themselves, and not to the phenomena on which they have been imposed.

In vector representations, an explicit distinction is made between the locations of spatial entities and the non-spatial attributes of these entities. As mentioned above, these two types of characteristic are frequently held in separate database structures, although some recent work (e.g. Bundock, 1987) has sought to remove this conventional distinction by database integration. One of the major turnkey GIS packages, Arc/Info, actually comprises Arc, a spatial database and manipulation package, and Info, a commercial DBMS (ESRI, 1993a). The independence of the two subsystems is further illustrated by the ability to make alternative software combinations such as Arc/Oracle (Healey, 1988). Other turnkey systems, such as Genamap, were designed to provide spatial-data manipulation power which may be built on to an existing relational DBMS (RDBMS) using structured query language (SQL) interfaces (Ingram and Phillips, 1987).

A confusing variety of terms exists for the basic spatial entities involved in vector representations. It is therefore necessary to begin by defining the terms used here, and to note alternatives which may be found in the literature. The basic entities are illustrated in Figure 6.1:

1 *point*: each (x, y) coordinate pair, the basis of all higher-order entities;
2 *line*: a straight-line feature joining two points;
3 *node*: the point defining the end of one or more segments;
4 *segment* (also referred to as a chain or link): a series of straight-line sections between two nodes;
5 *polygon* (area, parcel): an areal feature whose perimeter is defined by a series of enclosing segments and nodes.

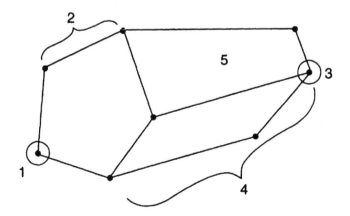

Figure 6.1 Basic spatial entities

In a classic paper on vector data structures, Peucker and Chrisman (1975) noted the extent to which data structure may be dominated by the ease of information input. More complex data structures require the input of topological in addition to coordinate information, and the task of reformatting or labelling spatial data which has been digitized with insufficient topological detail is extremely complex. Walker *et al.* (1986) use the term 'feature' for the fundamental geographic entity in the database, which has the following characteristics:

1 one or more geographic coordinates;
2 an optional associated text string;
3 optional graphic parameters;
4 symbol attributes;
5 optional user-defined attributes.

The simplest spatial entity in a vector representation is the point, encoded as a single (x, y) coordinate pair. In an efficient spatial database, each point will be recorded only once. The next level of entity is the segment, defined in terms of a series of points, and perhaps carrying additional topological information, such as the identities of polygons falling on either side. The third level of spatial entity is the polygon, for which a considerable variety of representation strategies exists. Due to the area-orientation of much socioeconomic data collection, data structures primarily relating to polygons have figured largely in the systems used for population data. Systems designed to handle data relating to the physical environment, such as utility-management and land-information systems (LIS) have tended to use more flexible structures which are better able to accommodate point and segment data types.

To illustrate the range of possible methods, we shall consider two representation structures which have been used for socioeconomic boundary data. These are illustrated in Figures 6.2 and 6.3, which show the same set of three zones encoded in different ways. The first of these is a polygon-based structure, which may be used for the storage of areal units such as census districts. In such a system, the polygon is the basic entity encoded, as shown in the boundary file in Figure 6.2. During digitizing into such a structure, each complete polygon boundary will be recorded as a single entity. This requires most points to be recorded more than once, as they will be contained in the boundary records of two or more adjacent polygons, such as points 1, 4, 5 and 7 in the figure. If digitizing is not precise, this requirement may lead to the creation of 'sliver polygons', one of the vector data errors noted in the previous chapter, where the coordinates recorded for a single point are slightly different in each polygon record. A key attribute such as the polygon identifier, is usually associated with each polygon. The inclusion of this item in the non-spatial attribute files allows rapid association of any attribute value with the polygon

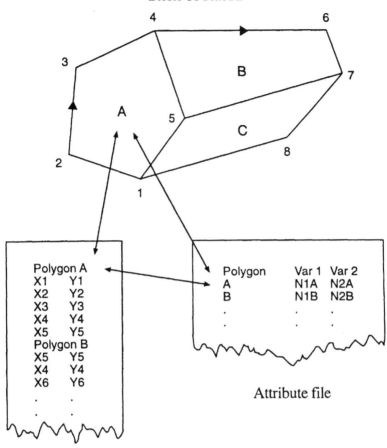

Figure 6.2 Polygon-based vector data structure

boundaries to which it relates. Polygon-based data structures allow rapid production of screen maps, as the data are already assembled into polygons suitable for display and area-shading subroutines. For this reason, polygon structures were particularly used in early mapping packages, and in PC-based software, where processing power for coordinate manipulation is limited. However, this simplicity for display operations is achieved at the expense of considerable duplication in the spatial database and, more seriously, with reduced flexibility for manipulation operations. Any changes to the spatial data must be made to every polygon record affected, and it is not possible to recombine the digitized points into a new set of areal units. A vector database, in the absence of any topological information, is effectively tied to the polygons originally recorded. Creation of new areal units, for example the definition of a new zone such as zone A + B in Figure 6.2, requires the recombination of segments belonging to

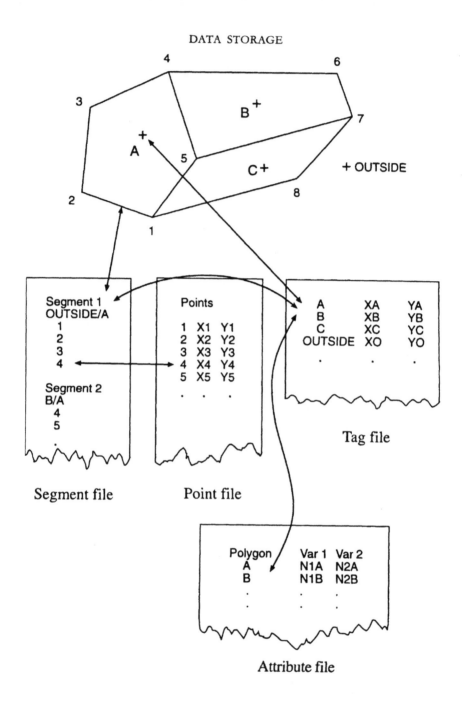

Figure 6.3 Segment-based vector data structure

different digitized polygons, and a polygon-based data structure does not allow for the individual identification of these segments. A structure of this type only requires one file for the storage of a map, as all the data is of the same type: lists of coordinates with associated polygon labels. Additional files may contain multiple attribute information for each polygon, indexed by a key attribute.

An alternative, and rather more advanced, representation strategy is to make the segments and points the basic entities of the database, with additional information that allows the reconstruction of the polygons at a later stage. Digitizing into this structure requires each segment to be identified and labelled separately in terms of the polygons which fall on either side. In Figure 6.3, segment 1 in the segment file is labelled as part of the boundary between polygon A and the 'outside' area of the map. Each point is stored only once, and given a unique point number. This information is stored as a separate file. A segment is stored as a pair of left/right labels, determined by the direction of digitizing, and a list of included points. The order of coordinates in the file will be that in which they were initially digitized, allowing unambiguous definition of the zones to 'left' and 'right'. These records constitute a second file of data. Thus the point reference numbers and their coordinates are recorded at separate locations in the database. In addition, a reference point may be recorded for each polygon, to facilitate faster display of area-based data and to aid in text label placement etc. These are known as 'tag points', and are stored in a 'tag file'. The attribute file is here identical to that used in the polygon-based structure in Figure 6.2. These points may also be used as an aid to segment labelling during digitizing (e.g. Vicars, 1986). A database of this type will thus typically consist of a number of linked files, each containing one aspect of the spatial data (point coordinates, segment records, area tag points). The advantages of this data structure include a more efficient database than that achieved with polygon encoding. It is possible to associate more than one set of left/right labels with each segment, as in the GBF/DIME file described on pp. 39–40. Also, different areal units may be constructed by processing zone labels: because each segment is a separate entity, new polygons may be assembled from any combination of segments distinguishable in terms of their polygon identifiers.

Taking UK Census zones as an example, the use of such a data structure would allow the extraction of boundary maps at ED, ward or district level. For example, a segment with left label DDAA01 and right label DDAB02 is identifiable as part of the boundary between EDs DDAA01 and DDAB02 and as between wards DDAA and DDAB, but it is internal to district DD, and would not need to be used in the construction of a district-level map (Figure 6.4). Attribute information is again linked by the use of key attribute fields, as in Figure 6.3, but in this case non-spatial data tables may exist for all definable areal units in the database (e.g. EDs, wards,

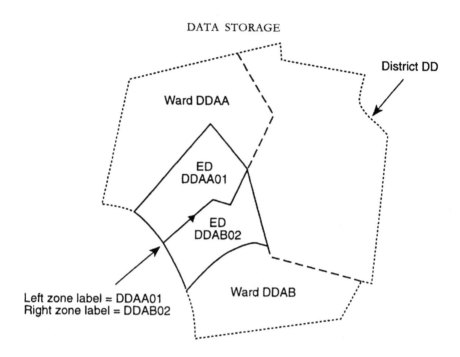

Figure 6.4 Census geography: segment labelling

districts), not just for the set of polygons identified. An important implication of the increased information in the point/segment/polygon structure is that two adjacent map sheets may be digitized, containing parts of the same polygons which fall across the sheet edges, and these polygons may be assembled from the data just like any other polygon. In a polygon-oriented system, a part-polygon of this type cannot exist, and either edge polygons must be included in both sheets, causing more duplication, or large overlaps are required between adjacent sheets to ensure that every polygon is wholly within at least one sheet.

The two structures outlined here are only two of a large number of possible vector data organization methods. An intermediate stage would be to make segments the basic units of the database, but to incorporate the point coordinates in the same file. This structure is still widely used as a low-level data-transfer format, as it can be read and interpreted by many different GIS packages. Many more complex structures are also possible. Additional information may be included in the spatial database according to the nature of the data. For example, unique segment labels such as road names may be required if the segments represent significant linear phenomena as well as describing zone boundaries; another example would be the inclusion of minimum enclosing rectangle (MER) information in segment and polygon records to speed spatial searches through the data. This comprises the highest and lowest values in x and y directions included in the current object (line, segment, polygon) and allows spatial searches to take place without the need to examine every point. Another example of computed information which may be stored with vector data would be the connectivity details required for network analysis. The inclusion of these extra pieces of information increases the size of the database but saves time in subsequent calculations.

Some general comments may be made about the data structures necessary for the storage of vector data. The basic difference between the representation types relates to the amount of additional topological information encoded with the coordinate data. This information will determine the manipulation operations which are possible on the completed database. While simple structures will suffice for basic map-drawing operations, or attribute calculations which do not involve the definition of new areal units, they are unable to support complex queries involving geometric computation or certain spatial criteria. The inclusion of more topological information demands more sophisticated (and potentially more efficient) data structures, and more processing for the basic operations. In order to draw one polygon from a point/segment/polygon database, a search of all segments is required to find those with labels matching the polygon required. These factors still have a significant impact on system performance where small computers or large databases are involved, but are not likely to be of enduring importance in the light of continued decreases in the cost of processing power. Vector-based structures are relatively compact, but high-quality data are essential. For this reason, considerable effort is required in the preparation of source documents and the verification and editing of digitized data before any analytical operations are performed. In the following chapter we shall consider some of the manipulation operations commonly used.

ATTRIBUTE DATA STORAGE

The storage of non-spatial attribute data is a well-established technology, quite apart from GIS. In its simplest form, it is analogous to a filing system, allowing information to be extracted from the database via some organizational structure. A traditional filing cabinet imposes an alphabetical structure on the data, allowing us to retrieve many pieces of related information from a record card indexed by an alphabetic name. We know where in the system to find objects referenced by a particular name, thus speeding up the search for information. For example, in an address record system, Mr Jones's address will be found on a card at location J, Mrs White's age will be located at W. This structure is illustrated in Figure 6.5(a).

In contrast, a computer-based database management system (DBMS) allows us to extract information, not only by name, but also according to a selection of the other pieces of information in each record: given an

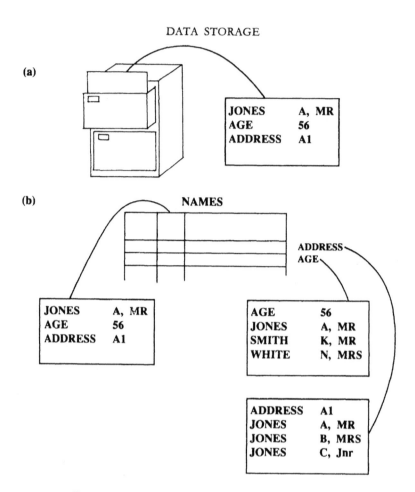

Figure 6.5 Manual and computer-based record systems

address, we could search to find out who lived there, or to identify all individuals with a given age, as shown in Figure 6.5(b). These operations would be impossibly tedious using a filing cabinet, which is only indexed for a single field of data (e.g. name of owner). As we have seen already, information systems are based around a digital model, which may be manipulated rapidly to perform the task required.

There are a range of organization strategies for DBMS packages, each of which imposes different models on the data, affecting the ways in which it can be extracted. The very simplest computer database would be a single-attribute file (as illustrated by the attribute file in Figure 6.3), with a number of fields of data relating to each object. From this, we may choose to extract the values of data fields for selected objects (as with the filing cabinet example) or search for values in other columns of the data, and find the objects to which they relate. Commercial systems use more sophisticated structures, allowing for complex relationships between objects. Date (1990) cites network, hierarchic, inverted list and relational structures as the four major categories. Software tools for the manipulation of data in these structures include methods for traversing the structure itself (e.g. finding all property owners with the same postal sector in their address, or finding all doctors serving patients on a particular register). The most popular structure for recent GIS software design has been the relational model (hence the term relational DBMS, or 'RDBMS'). An RDBMS consists of series of tables, linked by key fields, which allows the establishment of relations between the entries in different tables and imposes no navigational constraints on the user. The divisions between these tables are 'invisible' to the user, making this structure ideally suited to applications where the nature of queries is not known in advance. Whereas hierarchic structures, for example, embody rigid relationships between objects in the database, relational systems allow the establishment of temporary links between tables sharing common fields (Aronson, 1987). This serves to reduce redundancy and makes the database more flexible. The key fields in the geographic RDBMS are typically the unique identifiers of the point, line and area objects to which the attribute data in the tables relate. Healey (1991) notes the recent dominance of relational approaches, but stresses the advantages of hybrid approaches and object-orientation, considered below.

OBJECT-ORIENTED DATA STRUCTURES

Vector data structures make explicit distinctions between geographic entities in terms of their object classes (i.e. point, line, area), and are able to represent certain topological relationships between these objects. This may be contrasted with the field-based, or continuous, view taken in raster approaches. More sophisticated than either of these is the view of geographic entities taken by object-oriented approaches to data storage. In these, all entities are considered as belonging to one of a range of object types, and the approach is particularly well suited to the organization of geographic data which can be readily conceptualized as discrete objects. Clearly, this is well suited to the kinds of application conventionally associated with vector-based methods, but it is also possible to apply object-oriented ideas to raster-based data modelling.

Object-oriented approaches originated as a programming methodology, and the potential for efficient customization of existing software is much greater than with traditional programming languages. There has recently been significant growth in the use of object-oriented languages, such as C++, and a parallel development of object-oriented approaches in GIS, which is reviewed by Worboys (1994). As we have already seen in Chapter 4, phenomena in the real world are always represented by some form of data structure within a GIS. System implementation requires the development of operational data structures which mirror as closely as possible the characteristics of our conceptualization of these real phenomena. Object-oriented approaches allow the creation of objects which are much closer to real-world objects than the rather oversimplified notions of point, line and area. Object-orientation is concerned not only with the attribute characteristics which describe each object, but also with its dynamic behaviour. Objects with similar characteristics are grouped together into object types. At the implementation stage, there are developed for each object type data structures which contain both attributes and behaviour.

Each object type is thus defined as having certain properties (e.g. a land parcel may have properties of address, owner, value), and these types have associated operations which may be performed on them (Herring, 1987). These operations are invoked by the passing of 'messages' from one object to another. Topological information, including such concepts as adjacency, connectivity and orientation, may be used to describe the relations between entities. There is still some disagreement as to what precisely constitutes an object-oriented data model, and thus different variations emphasize different constructs. Many of these constructs, however, have useful applications in the context of geographic information (Aybet, 1994). Object identity allows objects to have unique identifiers independently from their attribute values, permitting the selection of features by a unique key or name. This also allows an object to retain its unique identity even if all its attributes are changed. Other constructs emphasize the re-use of object definitions, such as encapsulation and inheritance. Encapsulation is a programming concept which ensures that all sub-components behave consistently in any setting; thus an object may only be accessed by clearly defined procedures. In this way, information is 'hidden' inside the object and can only be accessed by using a predefined external interface. The construct of inheritance permits some object types to be defined as generalizations or specializations of existing types, as illustrated in Figure 6.6. For example, the generalization of

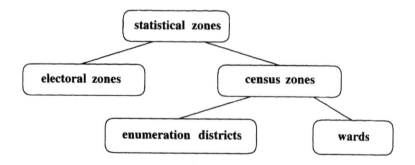

Figure 6.6 Object orientation: object inheritance

106

'census zone' and 'electoral district' object types might create a new 'statistical zone' type which inherits some of their characteristics. Similarly, 'enumeration district' will have the general properties of both statistical zones and census zones, such as area, population and designation date, in addition to certain unique attributes of its own. Complex objects may be created by aggregation or association of component objects. Polymorphism is a construct allowing the same message to be sent to different classes of object. Thus we may issue simple display commands which are interpreted correctly by both zonal and linear objects. Any special instructions for implementing the display of a particular object class, such as the need for shading or symbols, will be contained within its own definition. All these are powerful tools for the representation of complex geographic objects, and a particularly useful property of this general approach is the way in which error characteristics associated with an object may be incorporated in the spatial database (Guptill, 1989).

By avoiding the conventional vector separation of spatial and attribute data types and including location as one of the object's properties, the management of data input and output is much simplified. Many queries involving topology may be solved by reference to the relations inherent in the data structure, and coordinate processing is only necessary in very localized regions. Although the concepts discussed here may at first seem more closely allied to vector modes of representation, the object-oriented approach may also be used in conjunction with raster-based GIS, with the object record containing reference to particular cell codes in the raster coverage (Jungert, 1990). Queries using object-oriented structures tend to be very natural, but the form of such queries is more strongly predetermined by the organization of the data. At present, the database structures and query syntax suitable for object-oriented GIS are still under development and have not been standardized. A number of hybrid systems exist which use object-oriented language in conjunction with relational databases such as Oracle (Jianya, 1990). Worboys (1994) argues that for the future it will be necessary either to develop more sophisticated extensions to the relational model, which are able to handle both spatial and attribute information within a unified framework, or to adopt radically new approaches such as object-orientation. This idea is echoed by Boursier and Faiz (1993), who stress the lack of power of the relational model for handling spatial data, although they acknowledge that the potential of object-oriented systems remains to be proven. Such systems still account for a very small portion of the commercial GIS software market, although systems such as Smallworld GIS (Chance *et al.*, 1990) have made significant advances in this area. Many commentators predict that the importance of object-oriented approaches is going to increase (e.g. Maguire and Dangermond, 1994; Rix, 1994).

RASTER DATA STORAGE

Unlike vector representations, raster approaches do not encode the world by identification of separate spatial entities, but use thematic coverages. A data value is assigned to every cell (commonly referred to as a 'pixel' or picture element) of a georeferenced matrix, covering the entire data plane. Each coverage or layer in the database may thus have its own unique geography, independent of any boundary features except the edges of the pixels themselves. Point and linear entities may be encoded in such a data structure, but the result is a whole coverage which includes these elements, not individual database entries. As we have seen in Chapter 2, early developments in computer mapping were often restricted to crude raster-based output such as line-printer maps, and the view has become widespread that raster models are less accurate than vector ones. The key issue here is that of resolution: vector data may apparently offer greater precision than the encoding of a point or line in raster form, but there is no guarantee of greater accuracy, especially when the raster cellsize is small. Raster data with a cellsize of 1 m on the ground offer exactly the same locational precision as vector data with coordinates recorded to 1 m. In either case, the issue of accuracy is related to the quality of data collection and subsequent errors in processing: it is not directly a function of the decision to use either a vector or a raster data structure. Contemporary systems are capable of producing extremely high-quality raster graphics, and in addition there may be applications in which the raster model is more appropriate as a representation of the real-world phenomena.

A database may contain a very large number of coverages, in which spatial and attribute information are combined, geographic location being represented by position in the data matrix. Values stored in the pixels of a single layer may be dichotomous (present/not present), discrete classes, or continuous values. The simplicity of raster representations offers very fast manipulation and analysis, but this is achieved at the expense of storage efficiency. Raster coverages containing large areas of contiguous pixels with the same attribute value lead to massive redundancy in the database, and overcoming this redundancy has been a major issue in the development of raster data structures. A basic limitation of these techniques is that of pixel size. The pixel size determines the smallest data variation which can be recorded, and both pixel size and grid position may be very important, especially when the mapping unit (contiguous area of identical data value) is similar to the size of the pixels in the grid (Wehde, 1982). Accuracy inevitably decreases as pixel size increases, but smaller pixel size means more data values.

Zobrist (1979) itemizes the uses to which raster representations may be put and, before discussing the storage and manipulation of these data in detail, it is worth noting the types of variable which may be referred to by the values in the pixels:

1 *physical analog*: e.g. elevation, population density;
2 *district identification*: e.g. district which includes that pixel;
3 *class identification*: e.g. land-use or land-cover type identification schemes;
4 *tabular pointer*: e.g. record pointer to a tabular record (similar to the key-attribute field in a vector database and RDBMS);
5 *point identification*: presence of a point feature within the pixel;
6 *line identification*: presence of a linear feature passing through the pixel.

The simplicity of raster representations of geographic phenomena is both their strength and their weakness: the storage of some value for every location in space makes many manipulation operations very fast. There is no need to work out what is at a particular location: it is actually encoded in the database, but this comprehensive cover demands large volumes of data. As such, all raster representation methods are basically very similar, but great variety exists in the techniques used to achieve efficiency of storage in the digital data. In raw remotely sensed images, for example, every pixel may have unique combinations of values in each waveband (combinations of reflectance not repeated precisely in neighbouring pixels), but once the image has been classified, large numbers of adjacent pixels will share the same attribute values (e.g. a land-cover type). This is also the case with vector polygon maps which have been rasterized (see pp. 127–129) such as for administrative areas and soil and geology polygons. The object of an efficient data structure is to reduce the amount of redundancy in the database, i.e. to use as few records as possible for recording the location and value of each contiguous group of identically valued pixels. In the following descriptions, lines of pixels parallel to the x axis are referred to as 'rows' and those parallel to the y axis as 'columns'. Two different methods for compacting an image for storage are considered here, although in practice each of these has a number of derivatives commonly used in raster GIS, and there are a variety of alternatives, including the separation of pixel identifiers and multiple attribute tables, as used in vector systems (Kleiner and Brassel, 1986; Wang, 1986).

In a full raster matrix (Figure 6.7), every pixel requires one record, and many of these may be identical, as the data file is a direct model of the map cells. In Figures 6.7 to 6.9, the same simple map is encoded, with the black region having value 1 and the background having value 0. As multiple thematic overlays are produced, the unnecessary duplication of data values becomes unmanageable, and some alternative storage structure must be found. Even in image-processing (IP) systems, in which the value for each pixel is limited to a single integer number in the range 0-255 (eight binary digits), one screen image of information for one LANDSAT MSS band may occupy considerable amounts of disk space: for example, a 500 × 500 pixel

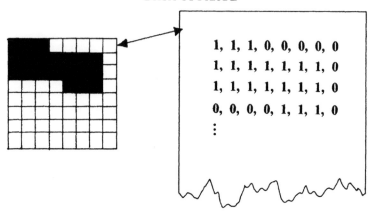

Figure 6.7 Full raster matrix

image will require 0.25 Mb per waveband. Many raster GIS systems use full matrix formats for data import, with the option to hold cell values as floating-point numbers (this being a method of storing decimal numbers in the computer). Frequently, however, data values are then classified and compacted for internal storage, in an attempt to reduce the disk storage requirements of the system.

If a coverage is very sparse (i.e. there are very few non-zero cells), as may occur with a map of point events or a population distribution, it may actually be more efficient to store the full row and column locations of each non-zero cell with its attribute score, rather than use a technique to encode the entire matrix. Most methods, however, seek to reduce the amount of redundancy in the data by identifying contiguous cells containing the same attribute value and grouping them together.

One alternative is to encode the image by examining each row in turn and identifying homogeneous runs of pixels. These may then be stored by recording their start and end positions and value, as illustrated in Figure 6.8. This approach is known as 'run-length encoding', and may dramatically reduce the number of records required to store an image. Some approaches allow for a run of pixels to continue from the end of one row to the beginning of the next, while others always clip at the end of a row. The idea of processing the image row-by-row still suffers from unnecessary duplication in the sense that homogeneous pixel groups are only identified in one direction (parallel to the x axis), whereas values in nearby rows (either above or below) are also likely to have the same values, but are represented separately. This feature of image-based data is exploited in the use of tree-based data structures, and these (especially quadtrees) have received much attention in the GIS literature (Mark, 1986; Laurini and Thompson, 1992).

Tree-based structures are assembled by the breakdown of the image into

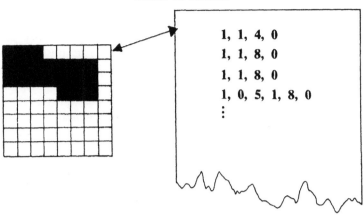

1, 1, 4, 0
1, 1, 8, 0
1, 1, 8, 0
1, 0, 5, 1, 8, 0
⋮

Figure 6.8 Run-length encoding

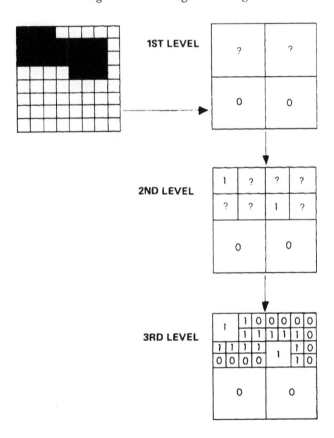

Figure 6.9 Quadtree data structure

111

successively smaller spatial partitions, until all pixels falling within each partition share a common attribute value. Breakdown of that partition then ceases, and a record is created which contains its position, level in the tree structure, and attribute value. In the case of a quadtree (Figure 6.9), the tree structure is of order four, and subdivision of the image is into quadrants. Figure 6.9 illustrates the breakdown of the simple region at three levels of the tree hierarchy: the whole image, represented by the outside square, requires further subdivision; the first division into quadrants reveals that the SW and SE quadrants are uniform regions of value 0. The NW and NE quadrants contain more than one attribute value and thus require further subdivision, indicated by a '?' in the figure. At the second level of decomposition, two quadrants are uniformly valued 1, but a third level is required to achieve a complete representation of the original map. In this example, at each level of the tree structure, a cell may take one of three possible values: complete presence (1 in the figure), complete absence (0 in the figure), or a mixture of both. So long as we step through the quadrants in a consistent order (e.g. NW, NE, SW, SE), it is only necessary to store one of these three values, as their location will be implicit from their position in the data structure. With reference to the diagram, we would begin by recording that the whole image is subdivided; then that the NW quadrant is subdivided (1st level); that the NW quadrant of this is entirely valued 1 (2nd level); the NE quadrant is further subdivided (2nd); within this, the NW quadrant is valued 1 (3rd), NE is 0 (3rd), SW is 1 (3rd) and so on. . . . When final values (0 or 1) have been encountered for all quadrants in a particular subdivision, we move on to the next quadrant at the level above, until the entire image has been encoded. The actual data values which we have represented as 'subdivision', '1' and '0' will vary between different software systems, but will generally be stored very efficiently. A number of different sequences may be used for stepping through the quadrants, including space-filling curves such as the Peano and Hilbert curves. These offer various advantages in indexing and identification of neighbours, and are discussed in more detail by Laurini and Thompson (1992). One of the greatest advantages of quadtrees is thus their variable resolution, within a single thematic coverage.

If one of these data compaction methods is used, additional processing is required to store and retrieve the original data matrix. Many manipulation operations will be affected by the ease with which they can access information about a particular pixel. Holroyd (1988) examines a variety of compacted structures in terms of storage efficiency and basic IP operations, and suggests that the simpler run-encoded structures may have significant processing advantages, except where complex transformations are required. The usefulness of these techniques rests therefore with the nature of the attribute being represented. 'If each cell represents a potentially different value, then the simple $N \times N$ array structure is difficult to improve upon' (Burrough, 1986). Mark (1986) notes that LANDSAT MSS images, digital elevation models (DEMs), and rasterized point and line files are rarely suitable for quadtree encoding.

It is apparent that, although essentially simple to encode and store, raster approaches have limitations in the representation of certain types of spatial phenomena. The storage of values for every location in space makes the database very large, but unless great care is taken with data generation techniques, the spatial units may still be too coarse in the areas of greatest significance, while remaining inefficient elsewhere. The use of regular spatial units does, however, offer fast and powerful geographic analysis, if assumptions about data validity can be met.

TRIANGULATED IRREGULAR NETWORKS

Before moving on to the discussion of some basic data-manipulation operations, a special technique for the representation of surfaces is worthy of a little more consideration. The main application for these models has been in the representation of the elevation of the land surface, known as digital elevation modelling. This technique is of interest here because a number of socioeconomic phenomena may be considered as continuously varying surfaces. An example of a study in which a triangulated irregular network (TIN) model has been used for a population density surface may be found in Sadler and Barnsley (1990).

The simplest way of storing a surface model is as an altitude matrix, which is a continuous-value raster map. In this instance, the raster model is being used to represent a surface entity which is in reality present at every (x, y) point in geographic space. As no data structure can contain an infinite number of real points, the surface must be represented by a finite number of locations in the data model (Yoeli, 1983), which is thus a generalization. It may be produced by various means, including interpolation from point height values and computation from other sources of digital data. DEMs illustrate some of the difficulties encountered with raster representations: although not ideally suited to quadtree encoding, there may be large amounts of redundancy in areas of uniform terrain, and the regular grid structure may distort certain operations such as line-of-sight calculation. The matrix may be too coarse to capture very localized features, yet their misrepresentation may lead to major problems in data analysis. Alternatively, too coarse an attribute encoding may fail to represent small changes in elevation (Ley, 1986). An alternative representation strategy has been developed, using a triangulated irregular network (Peucker et al., 1978). TIN models allow for extra information in areas of complex relief without the need for redundancy elsewhere in the representation. Figure 6.10 illustrates altitude-matrix and TIN representations of a simple cross-section of a surface. The resolution of the TIN adjusts to the amount of irregularity

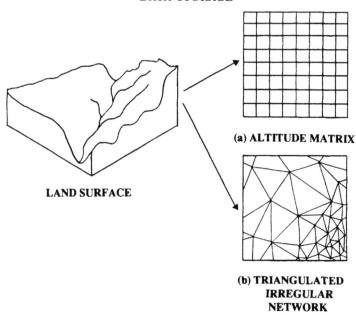

(a) ALTITUDE MATRIX

LAND SURFACE

**(b) TRIANGULATED
IRREGULAR
NETWORK**

Figure 6.10 Altitude-matrix and TIN methods for representing surface values

in the land surface. A key feature of TIN construction is the use of points of high information about the form of the surface (e.g. summits, points on ridge-lines and changes of slope). The triangulation of these data points is usually performed using a method known as Delaunay triangulation, which is described in more detail on pp. 133–135 (Tsai, 1993). TINs may be constructed from raster coverages to provide more efficient representations of the same data. It must be stressed that none of the structures discussed here is ideally suited to the representation of all types of surfaces.

SUMMARY

In this chapter we have considered various aspects of data storage. The database component of a GIS has been viewed as an abstracted model of reality which incorporates certain key parameters of actual objects. This model is constructed from the data collected and input using the techniques introduced in the previous chapter. Once verified and corrected, the data must be organized in a way which reflects the user's view of how reality is structured. There are some general principles which apply to all kinds of data model, such as the need for easy access and the elimination of redundancy due to inefficient structuring.

These aims may be achieved at two levels. The first, which has not been considered in detail here, concerns the way in which values are packed together, processed and stored within the mechanism of the computer. This level is rarely of concern to geographers, planners and other GIS users, except that it should work effectively and invisibly. The second level concerns the organization of the input data into objects or coverages to reflect the way in which we think about the world. If the user's concept of a real phenomenon is of an essentially areal entity, the most appropriate representation will be as an areal feature in the database. This will allow the representation of the object to be manipulated in the same ways as its real counterpart (i.e. area measurement, overlay operations, etc.). If, however, the phenomenon is most helpfully conceptualized as a continuous surface, an areal representation will be misleading and will not be amenable to surface-type manipulations.

A conventional distinction between GIS databases is between vector and raster modes of representation. However, object-oriented structures and triangulated irregular networks also offer means of representing real-world phenomena. There are increasingly common instances of integration between one or more of these structures, again suggesting that the distinctions are more apparent than real. In the light of contemporary technology, we must conclude that there is no simple answer to the question of which is the 'best' data structure, but that we need a clearer understanding of the implications of the techniques we are using.

The reason why these issues are so important is that the form of the model affects every aspect of the manipulation and analysis of the data. If the data are structured as areal units in some way, but the user views the underlying phenomenon as a continuous surface, it will not be possible to analyse the data in a way consistent with the user's concepts. In the next chapter, we shall look in much more detail at the techniques of data manipulation, the third transformation stage in the framework for GIS introduced in Chapter 4.

7

DATA MANIPULATION

OVERVIEW

The third transformation stage in our model of GIS as a spatial data processing system is that of manipulation and analysis. We have now seen how data about the world are collected and entered into the computer, and the ways in which they can be organized into a model of geographic reality. The very first manipulation operations are those concerned with the verification and correction of spatial data after input, and these were introduced in Chapter 5. If the process of data input were perfectly 'clean' and digitizer operators never grew tired or made mistakes (!), these operations would be unnecessary, so they have been separated from the main body of manipulation and analysis operations considered here. The majority of GIS software systems make the assumption that the data they are processing are entirely accurate, although this is rarely the case. There is an urgent need for users of GIS to understand the implications of the likely errors in their databases and for software which is able to take this into account to some extent (Veregin, 1989; Chrisman, 1991).

The ability of GIS to query and modify their model of the world is considered by most authors to be the 'heart' of GIS. For some, it is this ability which determines whether or not a system is truly a GIS, and for others the nature of these manipulations forms the basis for their theoretical model of GIS. This latter view has been characterized in Chapter 4 as the 'fundamental operations of GIS' model. Certainly, the manipulation capabilities of a fully functional GIS will be far more extensive than those of a computer-assisted cartography (CAC) system, whose main concern is with the display and presentation of map images, with only limited reprojection and editing facilities. There is increasing demand for sophisticated analytical procedures to be incorporated into GIS software, to make full use of the powerful geographic database, but these developments have been slow to appear in the commercial software systems (Goodchild *et al.*, 1992). Many early users of GIS-type systems were attracted by the display technology, but now find their software

116

unable to meet the demands of complex queries and analysis. An important feature of spatial databases is that they should be able to support analysis using truly spatial concepts such as adjacency, contiguity and neighbourhood functions. Many commercial systems offer powerful tools for data query and overlay, but have been much slower to develop statistical functions and a full range of data-conversion routines. Most of the statistical analysis functions which have been implemented use many of the same principles as the basic manipulation operations discussed here.

The procedure used for answering a query or performing an analysis on a spatial dataset is to a large extent dependent on the structure in which those data are stored. Therefore, the following sections treat vector and raster manipulation methods separately. In the same way, TIN and object-oriented data structures will require special procedures, but these are much less common than simple vector and raster approaches and are not considered in detail here. In many applications, raster manipulation is faster and easier to program than vector, as the entire data matrix may be manipulated as a two-dimensional array within a program. No coordinate processing is required, and geographic location is more easily computed from location in the database. However, some types of manipulation, particularly those involving reprojection, are much more complex than with vector structures.

In addition to these, some of the most important manipulations are those which are used to convert data between storage structures (e.g. vector to raster and raster to TIN), and those used to convert data from one class of spatial object to another (e.g. point to area and area to surface). These manipulations are of major significance because they create new geographic datasets which may be stored in the database and used for subsequent analyses. Consequently, they represent the user's main tool for altering the way in which the database models geographic reality.

VECTOR DATA MANIPULATION

Some specialized manipulation functions may be found in CAC and image processing (IP) systems (such as map reprojection, image classification), but it is the general, analytical manipulation operations on spatial data which most authors suggest identify a true GIS (e.g. Berry, 1987; Dueker, 1987). These operations have largely been developed to answer specific user queries. Extensive examples of these types of query are given in Young (1986), and are of the following types:

List the names and addresses of the owners of all land parcels falling wholly or partially within 500m of a projected road centreline.

117

Examine whether the cases of respiratory disease in a medical register are significantly clustered around any of the potential point sources of atmospheric pollutants in the neighbourhood.

Identify the settlements in a region having population characteristics likely to be able to support a new branch of a store, and identify the locations of potential competitors.

These manipulations may be divided into those involving only attribute calculation, those involving only spatial calculation, and those involving a combination of attribute and spatial (the classification of basic transformation operations used in Chapter 4). The solution of these questions may therefore require reference to both spatial and attribute databases, and may involve geometric and mathematical operations on the retrieved data (e.g. the calculation of a new polygon entity corresponding to a 500m buffer along a road centreline). Output may be in the form of attribute listings or new map information, or may be exported directly to another software system.

In some systems, the manipulation capabilities may include sophisticated analytical operations for certain datasets, such as point pattern analysis or network linear programming. Recently, the need to incorporate more general spatial-analytical methods within GIS frameworks has begun to receive more attention (Goodchild *et al.*, 1992). At the core of most manipulations which involve retrieval of spatial data is a relatively small set of operations which may be used many times, such as radial searches, tests for spatial coincidence, and the identification of all entities within a geographic window. Fortran source code for a variety of vector manipulations may be found in Yoeli (1982). These all involve a considerable amount of coordinate calculation and make use of the topological information encoded in the spatial database. Teng (1986) notes the need for 'topological intelligence', especially in determining the spatial relationship between any two sets of geometric elements, by overlaying digital map themes. The basic vector overlay operations are illustrated by way of a matrix of 'search' and 'target' entities in Figure 7.1. Overlay operations may involve any combination of point, line and area entities, and different mechanisms are required to perform each class of operation. Some of these are relatively simple, such as the calculation of the intersection of two line segments (cell 5 in the figure), but others such as polygon intersection (cell 9), or point-in-polygon searches (cell 3) are complex geometric calculations for which a variety of algorithmic solutions may be identified (Doytsher and Shmutter, 1986).

Brusegard and Menger (1989) outline a number of manipulation operations which are commonly performed on census-zone and postcode (point-referenced) data. Carstensen (1986) stresses the need for a good understanding of the nature of the spatial data if meaningful operations are

'TARGET' ENTITY

		point	line	area
	point	1	2	3
'SEARCH'	**line**	4	5	6
ENTITY	**area**	7	8	9

Figure 7.1 'Search' and 'target' entities in vector analysis

to be performed. For much socioeconomic data, the calculation of inter-section information for areal units is a meaningless operation, as the relationship between the data distribution and the areal units is unknown. In many such situations, it is necessary to estimate the values to be assigned to the new polygons formed as a result of overlay, and a variety of strategies exist. This problem is known as 'areal interpolation', and has recently received considerable attention in the GIS literature. It is discussed in the context of data interpolation on pp. 132–133.

Another situation commonly faced in socioeconomic analysis is the association of point (e.g. address-referenced) data with a particular areal unit for the calculation of, for example, incidence rates, and the steps necessary for this task are outlined here.

Given a list of current customers' socioeconomic characteristics, iden-tify addresses of potential customers having similar characteristics, as a target group for a mailshot.

The operations required in answering this query are illustrated in Figure 7.2. First, the customer characteristics may be matched to a neighbourhood 'type' in one of the commercially available classifications. This is a purely attribute-based operation, and may be performed in the attribute DBMS. There is then a need to examine the classifications of all areal units in the database to extract boundary coordinates of all polygons with the same classification. At this stage, a spatial operation must be performed: post-code locations (in the UK from the Central Postcode Directory, or in the USA derived from ZIP Plus 4 data) must be checked against the polygon boundaries retrieved in order to find those falling within each polygon. This is performed by using a point-in-polygon test. Alternatively, directory

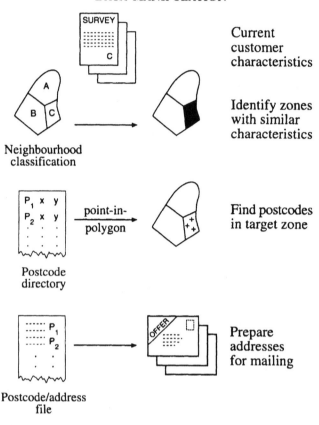

Figure 7.2 A sequence of vector data manipulations

information may be available which indicates the degree of overlap between postcodes and areal units, such as the 1991 directory of enumeration districts and postcodes in the UK (OPCS, 1992b).

The basic logic at work in a conventional point-in-polygon routine (Laurini and Thompson, 1992) is to assemble the segments associated with each polygon in turn, and to construct a line from the point to be tested to an edge of the map, counting the number of intersections with lines belonging to the current polygon (see example in Figure 7.3). The number of intersections returned will be an odd number only if the point falls within the polygon boundary, regardless of boundary irregularities. Clearly, such an operation is heavily dependent on the quality of the digital data. The existence of a single sliver polygon or a segment entered twice in the database would make the entire operation unreliable. The test also involves a relatively complex series of operations for each point. The use of polygon minimum-enclosing-rectangle (MER) information will allow for

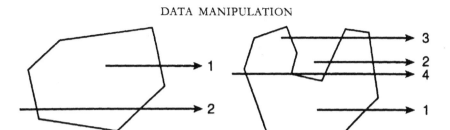

Figure 7.3 Point-in-polygon test

the rapid elimination of distant polygons from the search procedure. The focus then shifts to the individual lines, and it is necessary to identify every boundary line which falls across the test line. In most cases this can be done in terms of the minimum and maximum x and y values for each line, but in some cases will involve an intersection calculation on the two lines, using Pythagoras' theorem. Finally, addresses in each postcode may be obtained (e.g. from the UK Postcode Address File), and used to prepare material for mailing.

Having described this series of operations in some detail, it is necessary to add some comments. The assignment of postcode locations into each statistical zone need only be performed once, and will then be of use to a number of analyses. For this reason, commercial data service companies will perform these re-usable manipulations once and add the zone identifiers to their postcode-referenced databases, enabling them to offer rapid address-matching of clients' data at low additional cost. The purpose of describing this technique is to illustrate the complexity involved in even a relatively simple vector operation such as establishing whether or not a specified point falls within the boundary of a polygon. Such operations (point-in-polygon, area of a polygon, area of intersection between two polygons, etc.) are frequently encountered in the analysis of socioeconomic data. An alternative method for the assignment of postcode locations to data-collection zones would be to simply allocate each postcode to the zone having the nearest centroid location, where these are available. This involves less coordinate processing and does not require digitized zone boundaries, but the resulting allocations will be considerably less accurate, as examined by Gatrell (1989) in the context of UK postcodes and EDs.

Despite the geometric accuracy which may be achieved in answering such questions, the calculation required is time-consuming and complex, and in many cases may be either far in excess of the precision of the data itself (see for example the method of 100m grid reference assignment for unit postcodes on pp. 76–77), or totally unsuited to the nature of the data (such as the assumptions inherent in the use of some areal boundaries).

Users should be aware of the problems of very low-attribute counts when using small areas, as these may often invalidate traditional statistical analyses (Kennedy, 1989). Openshaw (1989) raises the issue of 'fuzziness'. More recently, a number of authors have begun to experiment with the use of fuzzy logic in a variety of GIS application fields, as a means of handling imprecision both in data and query definitions (e.g. Kollias and Voliotis, 1991; Banai, 1993). Clearly, errors may be introduced at any stage in the above process, or may already be present in the data. In the light of this knowledge, additional useful information may be obtained by finding postcodes falling in 'similar' neighbourhood classifications and for districts adjacent to those highlighted by the method described above. The identi- fication of adjacent zones will require analysis of right/left zone labels on the segments associated with each polygon – again, an explicitly spatial operation, which could not be performed by an attribute-only DBMS. These modifications are really attempts to utilize the inevitable error in the deterministic allocations of point-in-polygon.

Many spatial-analytical procedures have been developed which work with coordinate-based data, and these are increasingly being implemented within a GIS framework. For example, many of the techniques of point pattern analysis (Diggle, 1983) may be performed directly on a point dataset within a vector GIS. The GIS data model provides coordinate and attribute information in a well-structured way, and the basic system incorporates tools for the efficient retrieval of neighbouring points and the rapid calculation of distances between points. In practice, many of these innova- tions have been made in stand-alone software, which uses spatial data exported from a GIS, and which is re-imported for the purposes of display. Others have used the macro-level programming languages avail- able within the larger software systems in order to implement new statis- tical routines. This latter approach does indeed offer the scope for true integration of GIS and statistical-analytical tools, but is only available to the relatively small group of expert GIS users who are conversant with both the statistical procedures and the macro programming facilities of their GIS software. There remains little incentive to commercial GIS suppliers to include complex statistical functionality within their software when the need for such tools is not yet seen by the majority of their customers whose use of GIS is relatively routine.

RASTER DATA MANIPULATION

In the previous chapter, we considered the characteristics of the main data- storage strategies, and observed that raster representation is inherently simpler than vector, although sophisticated techniques may be used for reducing redundancy and overall size of the database. Once data have been unpacked and are in a matrix format, the manipulation algorithms required are able to operate directly on these structures.

The manipulation of raster databases differs from that of vector data in that it is not usually possible to directly identify spatial entities (such as field A or road B), and operate on these objects alone. The smallest addressable unit is the set of pixels contained within a single thematic coverage which share a common attribute value (e.g. all fields or all roads), or the pixels within a specified region of a coverage. In order to extract only these pixels, it is necessary to test every pixel in the image, so all operations effectively involve whole coverages. It is of course possible for the user to specify an individual pixel for processing in terms of its row and column position (e.g. by selection with a screen cursor), but this requires that the user have additional knowledge about the significance of individual pixels in the image. Manipulation takes place by arithmetic operations on the elements of the data matrix (each represented by a pixel in the image).

Although each image is georeferenced, individual pixels are only addressed in terms of their row and column positions within the matrix, hence there is no coordinate calculation involved in most coverage-based operations. This massively reduces the complexity of manipulation processes in comparison with the equivalent vector techniques. The absence of explicit coordinate and attribute information can be a problem, however. Even though visually identifiable features may be present in the pattern of pixel values, these cannot always be addressed directly by the processing system. This problem was noted in the classification of RS images by software on pp. 24–25. Devereux (1986) observes that the image does not contain all the necessary information to define a map: a certain degree of cartographic interpretation is also required. It should be stressed that the geometric correction and transformation (e.g. reprojection by 'rubber sheeting') of image data, and the creation of a new image with different pixel size by resampling, are very complex procedures. Although these are necessary tools in IP systems, they are still not found in all raster GIS software.

Despite these limitations, the great power of raster manipulation rests in the encoding of attribute values for the entire geographic space. Thus, the spatial relationships between real-world phenomena are maintained within the structure of the data matrix. For example, retrieval of attribute values surrounding a specified location merely involves the identification of adjacent pixels in the matrix by use of array pointers. No search of the entire database is necessary, and no coordinate distance calculations are required. To perform the equivalent of the point-in-polygon search, outlined on pp. 120–121 above, would involve overlay of the relevant point and polygon coverages in the database and a count of pixels with the required point value coincident with pixels having each polygon value. It is apparent that all points may be allocated to their corresponding polygon in one simultaneous pass through the two data matrices. The points and polygons cannot be treated individually, as they have no separate existence within the data structure. Only by use of a corresponding layer containing pointers to a multiple-attribute table would it be possible to obtain textual or un-coded information about any particular point or polygon. These examples serve to illustrate some of the key characteristics of raster manipulation operations:

1 Processing of an entire coverage is very fast, usually involving at most a single pass through the rows and columns of the data matrix.
2 Conceptually, any number of different thematic coverages may be involved in a single operation, the corresponding pixels of each coverage being processed simultaneously.
3 The results of most manipulation operations can be expressed in terms of a new map coverage.

The speed of these operations, and the creation of a new map at each stage, have facilitated the evolution of 'cartographic modelling' techniques (Tomlin and Berry, 1979; Tomlin, 1991), in which series of such operations are combined in a particular modelling scenario, each stage taking as input the result of the previous manipulation. The software routines forming the basic tools of such a system are essentially quite simple, and may be addressed using natural-language commands operating on specific coverages, such as DIVIDE, MULTIPLY, SPREAD, RECLASS, COUNT, etc.

A simple example of such a scenario is given here. Other examples, drawn primarily from the physical environment, may be found in Berry (1982) and Burrough (1986). Tomlin (1990) gives a comprehensive introduction to the role of cartographic modelling in GIS. Each operation is performed on the corresponding cells in each layer, so that in expressing map A as a percentage of map B to create map C, cell (1, 1) in A will be divided by cell (1, 1) in B and multiplied by 100 to produce (1, 1) in C, and so on, as illustrated in Figure 7.4. For this example, we shall assume the existence of raster layers containing district boundaries and also counts of population, unemployment, households, and heads of households aged 20–30. It is required to find the total population in each district living in neighbourhood type 'A', where type A is defined as having less than 5 per cent unemployment and more than 30 per cent of household heads aged 20–30. (This represents an extremely simplified example of neighbourhood classification!)

The modelling process is outlined in Figure 7.5, and might proceed as outlined here. In reality, each map would be a raster coverage, with values of the variables associated with each cell in the matrix. First, it is necessary to obtain percentage coverages for unemployment ('Unp'd') and household heads aged 20–30 ('hoh 20s'), by DIVIDE-ing the relevant variables by their base counts, and MULTIPLY-ing by 100. In some systems the exact form of these commands will vary, the first stage being performed by a

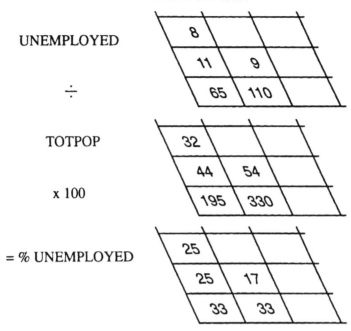

UNEMPLOYED

$\div$

TOTPOP

x 100

= % UNEMPLOYED

Figure 7.4 Raster data manipulation

single COMPUTE command, for example. The areas comprising neighbourhood type A may now be obtained by RECLASS-ifying the percentage maps, coding areas with the required combination of values as 1 and all other areas as 0. If some degree of locational fuzziness were to be included, as mentioned above, this classification could be SPREAD, by an extra cell width, and the additional buffer, representing areas adjacent to type-A neighbourhoods, given a code of 2. The total population may now be RECLASS-ified to contain 0 in all cells outside neighbourhood-A areas, and finally the resulting population in each district may be COUNT-ed, using either the precise or fuzzy definition of type A.

This example shows how the basic operations may be combined to answer highly specific enquiries. The operations outlined above, involving polygon overlay, buffering and attribute calculation at each stage would require a powerful vector GIS, and may still be prevented by the absence of a suitable data model (e.g. the need to estimate population in the fuzzy zone produced by polygon overlay). However, these operations are well within the capacity of even a relatively simple raster system, and the data requirements are less restrictive. It should be noted that most of the above operations will result in the generation of a new spatially coincident raster coverage and that the data volume created by such modelling increases rapidly with the number of steps to be performed.

125

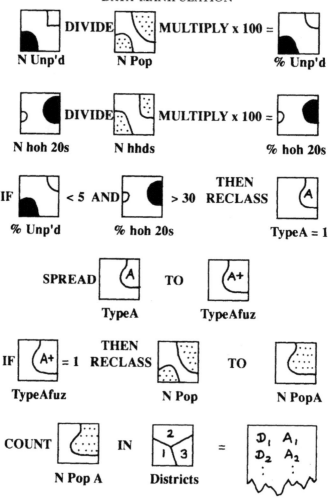

Figure 7.5 A cartographic modelling scenario

The fundamental difficulties with these techniques are not technical ones relating to the software for manipulation, or even the spatial resolution of the data, but rest more with the suitability of the data for modelling in this way. In a situation in which manipulation is so easy, data quality is crucial, and this relates especially to the methods used for data collection, generation and storage. The imposition of a fixed range of integer numeric codes on certain variables such as altitude matrices may make the use of such data both inconvenient and inaccurate. There is also a more general question as to what extent it is appropriate to use such deterministic 'map algebra', even assuming reliability in the input data. To some extent these issues are

126

addressed by the use of fuzzy concepts, but truly fuzzy thinking is at present rarely encountered in practice. It is now necessary to consider briefly the methods which may be used to transform data between raster and vector structures, and to examine particularly the types of interpolation technique which may be used in the generation of data for locations at which no data measurement is possible. These methods are more commonly used for the generation of raster structures, which require some data value for all locations in space.

DATA CONVERSION

Despite the long period during which raster- and vector-based spatial data handling systems have coexisted, there are still very few truly 'integrated' systems, which are able to handle data in either format with equal ease and to relate spatially coincident data stored according to the two different structures. There is still a need to convert data between representation types for specific operations, and no ideal techniques for these conversions have been discovered. Logan and Bryant (1987) describe the methods adopted for the routine transfer of data between the IBIS raster system and an Intergraph vector computer-assisted mapping system. Their conclusions are that the long-term solution must be an integrated system, but this is too far into the future for any operational GIS installation to consider, and the short-term solution is to develop efficient data-conversion software. Conversion is also commonly required when vector data such as contour maps have been digitized with a high resolution raster scanner and it is necessary to enter the data to the vector database (Weibel and Heller, 1991). Lemmens (1990) suggests three possible levels of integration. The first, and most elementary, is when raster and vector data may be overlaid visually and the relationships between the two datasets compared by eye. The second, with which we are primarily concerned here, involves a one-way flow of information from one dataset to the other. The third level of integration, which most contemporary systems have failed to achieve, is a two-way flow of information between the different data structures, ideally without operator intervention. The most promising advances in this direction have so far developed out of those image-processing applications in which ancillary data are used in image interpretation. Despite a number of research advances in this area, commercially available software still fails to deliver truly integrated vector and raster processing.

'Vectorizing' is fundamentally more difficult than 'rasterizing', as there is a need to create output data with a degree of precision which is not present in the raster source data. Additionally, there is a need for some type of topological information to be constructed and for individual features to be identified. This difficulty may be seen in Figure 7.6(a). The conversion of a

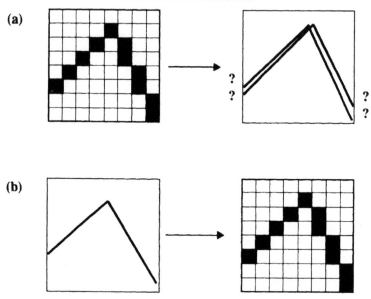

Figure 7.6 Vectorization and rasterization

raster database layer into vector format requires the identification and extraction of geographic entities from a structure which cannot represent these entities explicitly. Jackson and Woodsford (1991) note that this can only really be achieved where the finest lines are of at least 2-3 pixels width in the original raster representation. This is another problem similar to that of image interpretation encountered in the classification of RS data noted above. In the same way that IP systems may seek to identify edge features by detecting strong contrasts in the DN values present in adjacent cells, vectorization techniques examine neighbouring pixels in order to identify boundary features which may be significant to the vector database. The problem is a very complex one, and there is a lack of efficient algorithms, making the procedure very expensive in terms of processor time. Speed, output quality and accuracy are all important considerations, and none of the algorithms considered by Piwowar *et al.* (1990) were able to adequately smooth the 'stepped' nature of the vectorized edges. All such techniques are only really best estimates, and the most satisfactory results are usually achieved with simple datasets, which is why contours are one of the few coverages recommended for digitization by raster scanner when vector data are eventually required. The three main approaches are described in Peuquet (1981a):

1 *Skeletonization*: once linear features have been tentatively identified, less distinct pixel values at the edges of the features are successively removed

until the features are only one pixel in width; coordinate values are then derived from the centres of the constituent pixels.

2 *Line extraction*: once some part of a linear feature is identified, adjacent pixels are examined in an attempt to trace the route of the entire feature.

3 *Topology reconstruction*: in this approach, the entire image is analysed, looking for probable nodes and identifying their form (e.g. X, Y or T). The separate locations are then joined by the construction of lines between them.

Clearly, none of these techniques is able to reconstruct a full topologically structured vector database without considerable operator intervention and subsequent processing. The extraction of vector entities from raster images remains a significant obstacle in GIS capabilities.

Conversion of data in the other direction (Figure 7.6(b)) is a much simpler process, as not all the available topological information needs to be encoded in the raster image. Different raster layers may be created from the same vector spatial-data file by use of different variables from a multiple-attribute file. Two basic strategies exist: either the vector data may be processed entity-by-entity, assigning corresponding pixel values as they are encountered, or the image may be constructed pixel-by-pixel. There is a tradeoff between the amount of information to be held in the computer's memory and the speed of the process (Peuquet, 1981b). In polygon rasterization, pixel value assignment may be based on various criteria according to the nature of the attribute data. Possible criteria include assignment of a pixel to the polygon in which the mid-point of the pixel falls, or the calculation of the proportional contribution of different polygons to the area of the pixel and the calculation of a new weighted attribute value. Linear features are represented by the pixels through which they pass, and points by the pixels into which they fall. Nevertheless, this conversion operation is not without its difficulties: van der Knaap (1992) compares the performance of eight leading GIS systems in the rasterization of a specially prepared test map, revealing that the output differs greatly according to the manipulation algorithms used by the software designers. Carver and Brunsdon (1994) describe an empirical study using simulated coastlines, from which they examine the error processes associated with feature complexity, raster resolution and rasterizing error in greater depth.

DATA INTERPOLATION

With the possible exception of remote sensing, it is not usually possible to measure a geographic phenomenon at all points in space, and some sampling strategy must be adopted. Many different approaches to spatial sampling exist, according to the nature of the specific task. Comprehensiveness entails higher time and cost penalties, and may be inefficient in regions of little or no change over space. Variations on random, systematic or stratified sampling techniques may be used, according to the distribution of the phenomenon to be measured (Berry and Baker, 1968; Goodchild, 1984). The whole process of data collection, discussed in Chapter 5, especially the selection of sampling points, is thus crucial to all subsequent use of the measured data within any automated system.

Complete enumeration is possible with discrete entities such as individual buildings, but is impractical with a continuously present phenomenon such as land elevation or soil type. Exactly which objects in geographic space are considered to be discrete and which are continuous will depend to a large extent on the scale of analysis and the user's concept of the phenomenon being represented. Individual entities with precisely defined boundaries may best be represented by storage as distinct features in a vector database. Continuous phenomena may be represented in either vector or raster (contours, TIN or altitude matrix) structures, but either approach involves some interpolation of values for unvisited sites (Shepard, 1984). MacEachran and Davidson (1987) identify five factors significant to the accuracy of continuous surface mapping:

1 data-measurement accuracy
2 control-point density
3 spatial distribution of data-collection points
4 intermediate value estimation
5 spatial variability of the surface represented.

It is the methods for intermediate-value estimation (4), i.e. spatial interpolation, which are the focus of this section.

In conventional cartography, isolines (lines of equal value) may be interpolated by eye directly from the irregular input points. Automated interpolation is conventionally a two-stage process, involving initial interpolation of variables to a regular grid prior to map construction (Morrison, 1974). The basic rationale behind spatial interpolation methods is the observation that close points are more likely than distant ones to have similar attribute values. The aim is therefore to examine the form of data variation and find an appropriate model for the interpretation of values at measured locations to estimate values at unvisited sites.

A conventional distinction between interpolation methods is between 'global' and 'local' approaches (e.g. Oakes and Thrift, 1975). Global techniques seek to fit a surface model using all known data points simultaneously, whereas local interpolators focus on specific regions of the data plane at a time. This classification is used by some (e.g. Burrough, 1986) as an organizing concept for discussion. Here, however, the classification drawn up by Lam (1983) is preferred, as this focuses on the nature of the spatial data being interpolated rather than on the techniques them-

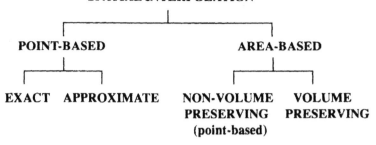

Figure 7.7 A classification of interpolation methods

selves. The general form of this classification is illustrated in Figure 7.7. Point methods refer to data which may be measured at a single point location such as elevation or rainfall. Areal methods relate to data which are aggregated over some areal units, such as census population counts. Point-based methods may be further subdivided into exact and approximate approaches. Exact approaches ensure that the output values for measured locations are precisely the same as the input data. Approximate approaches seek to minimize overall levels of error, and are not constrained to preserve all input data point values unchanged. Areal interpolation methods may be point-based, in which a single point is used to represent each areal unit, and one of the point interpolation techniques is then used. Such methods do not preserve the correct volume under the surface. Alternatively, the areal units themselves may be used as the basis for interpolation, in which case digital boundary data are required, and these methods do correctly preserve the total volume. An understanding of these principles is important here because of the significant role these procedures play in the construction of databases for GIS.

Point-based interpolation

Exact point-based methods include most distance weighting methods. The object here is to assign more weight to nearby data points than distant ones, The interpolator is exact when distance weightings such as $w = d^{-1}$ are used (where w represents the weighting value and d the distance). Such methods are, however, easily affected by uneven distributions of data points and are particularly susceptible to clustering in the input data. Weighted average methods are basically smoothing functions and may therefore fail to reproduce significant peaks and pits within a surface. These techniques do, however, have the advantages of simplicity and speed of computation and have been very widely used (Tobler and Kennedy, 1985). An alternative exact method is the fitting of a polynomial function of the lowest order which will pass through all data points. However, unreasonable values may be produced away from data points, and it may be possible to find more than one solution for the same data. Piecewise fitting presents problems of discontinuities in the resultant surface. More sophisticated techniques falling under the same classification include kriging, or optimal interpolation (Hawkins and Cressie, 1984). The technique is based on the recognition that the variable may be too complex to be modelled by a smooth mathematical function, yet still have some spatial dependence between sample values, which decreases with increasing distance apart. Kriging rests on the assumptions of regionalized variable theory, that the spatial variation of a variable can be expressed in terms of a structural component, a random spatially correlated component, and a random noise or error term. This technique was originally developed for use in the mining industry, where the most detailed estimates and their errors are required, but may impose a very heavy computing load in the process of surface generation.

Approximate methods involve the definition of a function which contains some residual value at every point. Various criteria may be employed to minimize the overall error. These approaches include trend-surface models. The use of such techniques really requires a strong theoretical justification, which is often not present in geographic applications. The distinction between the global trend and local random-noise components is often as much determined by the scale of analysis as by theoretical considerations.

Area-based interpolation

The areal interpolation problem is frequently encountered with aggregate data. Applications have tended to focus on the issues of isopleth mapping and the transformation of data between different sets of areal units (Flowerdew and Openshaw, 1987). This problem has received considerable attention as the use of GIS for the handling of socioeconomic data has grown. Reviews of the available techniques are provided by Goodchild *et al.* (1993a), and Moxey and Allanson (1994). The simplest approach is to weight the attribute values by polygon area, but this will only produce accurate results if the variable being estimated is uniformly distributed over space. Flowerdew and Green (1991) suggest a method for 'intelligent' areal interpolation, using the distribution of other variables in the database to predict target polygon values of the variable being estimated. In their example, data on car ownership and voting behaviour are used in order to assist in the interpolation of population data between two sets of zonal units. Goodchild *et al.* (1993a) illustrate an alternative method which uses information for control zones at different scales, and which may be applied where suitable ancillary data are not available. The precise form of any such estimation will vary according to the level of measurement (nominal, ordinal, etc) of the variables. The quality of the interpolated data is of particular importance if these are then used as the basis for subsequent manipulation and analysis.

Kennedy and Tobler (1983) address the related problem of interpolating missing values in a set of areal data. The traditional approach to these situations has involved the use of the point interpolators noted above, with a single reference point for each source zone, from which values are interpolated to a regular grid. The choice of suitable reference points for zones may in itself be problematic, as the distribution (before aggregation) of the variable within the zone is usually unknown and zone boundaries themselves may be highly irregular. Point-based interpolation of areal data also incurs the disadvantages of whichever interpolator is used, even assuming that the point location is meaningful. The final result will always be dependent on the method of area aggregation employed. The major weakness of all such approaches is their failure to preserve the total volume under the surface, the actual meaning of which will depend on the variable being interpolated.

Two volume-preserving methods are available, but both of these require digital boundary information, which is another time-consuming and costly constraint, if it is not already available. The overlay method superimposes target zones (which may be the cells of a regular grid) on source zones, and estimates target-zone values from the sizes of the overlapping regions. This is in itself one of the powerful manipulation capabilities of GIS, and is intuitively simple, although involving considerable coordinate calculation. An alternative method, pycnophylactic interpolation, has been suggested by Tobler (1979). This assumes the existence of a smooth density function $z(x, y)$, in which $z(x, y) = H_i / A_i$, where H_i is the observation in zone i, and A_i is the area of zone i. The interpolation begins by overlaying a fine grid of points on the source zones and assigning the average density value to each point falling within a zone. These values are then adjusted iteratively, by a smoothing function and a volume-preserving constraint, until there is no significant change in the grid values between each iteration.

Boundary interpolation

A different strategy in interpolation is the use of boundaries to define regions of homogeneity. These may follow some existing landscape feature or may be generated from data point locations such that an area is created around each point. A useful technique in this context is the construction of Thiessen (or Voroni) polygons. The complete subdivision of a plane into these polygons is known as a Dirichlet tessellation. The plane is divided in such a way that every location falls within the polygon constructed around its nearest data point, as illustrated in Figure 7.8(a). A technique for the

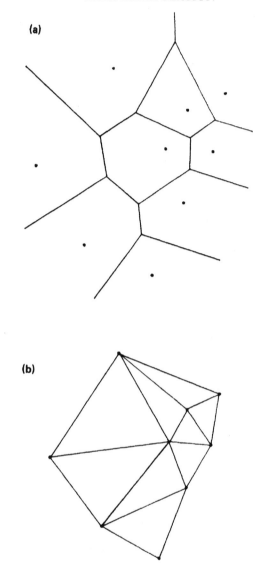

Figure 7.8 Dirichlet tessellation and Delaunay triangulation

computation of a Dirichlet tessellation is given in Green and Sibson (1978), and the concepts involved are discussed at some length in Boots (1986). A triangulation of the data points, termed a 'Delaunay triangulation', is used in the construction of the tessellation, and this may in itself be useful in providing ancillary information for various interpolation algorithms. The Delaunay triangulation joins data points whose polygons share a common facet, thus excluding data points which are masked by other, closer points, as may occur with data point clustering (Figure 7.8(b) illustrates the Delaunay triangulation for the set of points shown in Figure 7.8(a).) This triangulation forms the basis for the TIN-based DEM structure (Weibel and Heller, 1991). Sibson (1981) describes a method for 'natural neighbour' interpolation. Drawbacks of the Dirichlet tessellation for interpolation at unvisited sites include the fact that the value at each site is determined by a single data point value. The technique may best be suited to nominal data, where weighted interpolation methods cannot be readily applied.

Surface generation from centroid data

In the examples of cartographic modelling above, population and other census-derived surfaces were used. These may be generated directly from zone centroid data and offer considerable advantages over conventional vector representations. The technique described here was introduced in Martin (1989) and Bracken and Martin (1989). Here, a surface is generated in the form of an altitude matrix, where the 'altitude' in any cell represents the value of some socioeconomic attribute at that location.

The basic data requirement of the surface generation technique is a set of point locations which represent population-weighted centroids of small zones, for which count data are available. The boundaries of these zones are not required. This technique was originally developed in the context of the UK Census enumeration district (ED) centroids, but has subsequently been successfully applied to census data from other countries and to postcode locations in the UK. The underlying assumption of this technique is that the distribution of these locations is a summary of the distribution of the phenomena to be modelled. Population (or population-derived) counts associated with each data point are redistributed into the cells of a matrix, which forms the output surface-altitude matrix.

The generation of a surface from such a set of centroid information involves the use of a moving window, or kernel (Bowman, 1985), similar to the neighbourhood functions commonly found in raster GIS. The size of this kernel varies according to the local density of the centroids. The kernel is centred over each centroid in turn, and an analysis is performed of local inter-centroid distances in order to estimate the size of the areal units which they represent. The count (e.g. total population or persons unemployed) associated with each centroid is then distributed into the surrounding region, according to a distance decay function, with finite extent determined by the inter-centroid distances. Each cell of the output matrix falling within the kernel receives a weighting representing its probability of receiving a share of the count to be distributed from the current centroid. These weightings are then re-scaled to sum to unity so that the values given to the cells surrounding a data point exactly match its total population count. Where centroids are close together, the kernel will be small, and the count associated with each centroid will be distributed into a small number of cells, and in the extreme case may all be assigned to the cell containing that centroid. Alternatively, where centroids are widely dispersed, the count associated with each centroid will be spread across a greater number of cells, up to a predefined maximum kernel size. Distance decay functions, such as that introduced by Cressman (1959), which incorporate such a finite size, are appropriate for use in this context.

Ancillary information may be incorporated into this procedure: for example, a dichotomous raster map may be used as a mask to prevent the assignment of any count to certain cells. This may be useful to restrict the region for which the estimation is performed or to protect regions known a priori to contain no population (e.g. the sea). Once the count associated with a centroid has been redistributed into the surrounding cells, the kernel is moved on to the next centroid. When all the centroids have been processed, many cells in the output grid will have received a share of the counts from a number of different points, but in a typical region the majority will remain unvisited and thus contain a population estimate of zero. In this way, the settlement geography is reconstructed in detail from the distribution of the centroid locations even within the zones used for data collection. The cells each contain a population estimate, and the grid thus represents a height matrix of density of the phenomenon represented by the count values. The total volume under this surface (the sum of values in every cell) will be the same as the sum of the counts at the individual points. This volume preservation is an important characteristic as it enables the estimation of counts for ad hoc regions superimposed on the surface model. Centroids are processed which represent a slightly larger area than the region to be modelled, as this enables the construction of surfaces with no edge effects. Surface representations of adjacent regions may then be joined together to create a truly 'seamless' database.

The construction of databases covering large areas is straightforward, involving no more than the assembly of adjacent surfaces into a single file structure. Bracken and Martin (1995) and Martin and Bracken (1994) describe the creation of a national online dataset for the UK in this way, incorporating information from both the 1981 and 1991 Censuses of Population. In most regions, only a small proportion of the cells in a coverage will actually contain any population count. A consequence of this is that large geographic regions may be represented by relatively small data files. Even the densely settled region represented in Plate 5 contains a large proportion of unpopulated cells for which no information needs to be stored.

Once a surface of total population has been constructed for any area, this will form the basis for the calculation of all population-derived surfaces, such as age or unemployment. These surfaces are generated using count data for each variable to be incorporated in the database. Plate 5 illustrates such a population-density surface for a 100 × 100km region of Greater London, and Plate 6 is an enlarged portion of the same image, showing the grid-based nature of the model. The surface was derived from 1991 Census Small Area Statistics without the use of any digitized boundary data, and was extracted from the national dataset noted above. The shading values indicate the population estimates in each cell, with unpopulated cells in black and the most densely populated (1,800 people) in green. The map clearly shows the concentration of population in the areas immediately surrounding the metropolitan core, with relatively low values in the actual centre, which comprises mainly business and commercial land uses. Unpopulated regions are clearly identifiable, such as the Rivers Thames and Lea which are the major linear features running broadly west–east and north–south respectively. This type of model may usefully be compared with more conventional zone-based maps such as that illustrated in Plate 1. Displays of this type give a readily understandable representation of the geography of the modelled variable, reconstructing the locations both of unpopulated regions within the urban structure and of discrete settlements which are not separately identifiable from the zonal data. An alternative base model for the same region could be generated from household counts, which are also available in the Census SAS.

SUMMARY

In this chapter, three main types of manipulation operation have been considered: manipulation of data in vector structures, manipulation of data in raster structures, and the conversion or interpolation of data between different storage structures and object types. This ability to modify the model of geographic reality, using set of a widely applicable tools, is one of the key features of GIS, and distinguishes such systems from computer-assisted cartography and image processing.

These manipulations form the third transformation stage through which data in GIS may pass. The suitability of data for complex modelling and analysis will depend to a large extent on the error introduced by the earlier transformations of data collection and input (Chapter 5), and on the structures in which they are held (Chapter 6). The manipulation tools which may be employed are extremely diverse, including specialized algorithms for dealing with shortest paths around networks, and landscape modelling with TINs, for example. Just as no single system will incorporate all possible functions, it has not been possible to explain the working of all operations here, and therefore a selection of commonly required manipulations on socioeconomic data have been illustrated. As with the software systems themselves, there has been considerable development of deterministic modelling techniques in relation to the physical environment which need modification for application to data about human populations. As stressed in Chapter 4, the user's concept of the phenomena being modelled will often determine the most appropriate data structures and techniques to be used.

GIS databases potentially provide an ideal environment for the implementation of new and existing statistical analyses. These analyses operate on the same data structures and use the same geometric and logical operators as the manipulation functions, but have a very different aim. Openshaw (1991b) very clearly draws a distinction between manipulation and analysis, and identifies the origins of spatial analysis in the quantitative developments in geography in the 1950s and 1960s. Openshaw cites a range of analyses which are available for the investigation of data in each of point, line, area and surface types. Examples include nearest-neighbour analysis for a point datasets, network analysis of linear data, spatial interaction modelling of area data, and Bayesian mapping of surface data. The existing range of techniques contains relatively few exploratory tools. Many researchers have addressed themselves to the incorporation of such procedures within GIS and have variously identified sets of 'core' techniques, such as those tabulated by Goodchild *et al.* (1992).

It is possible to convert data between vector and raster structures without affecting the object class of the representation. For example, linear features in a vector database may be rasterized and encoded as linear entities in a raster coverage, and vice versa. However, the conversion of data from raster to vector remains problematic, due to the need to construct topological information and locational precision which are not present in the input raster data. Truly integrated systems which facilitate two-way information flow between raster and vector databases are not yet widely available. Many large systems contain separate modules for the handling of data in different structures, together with a selection of routines for the conversion of data from one structure to another.

Alternatively, techniques exist for the interpolation of data from one object class to another, such as points to surface or points to areas. These methods are of particular interest in the socioeconomic context, as it is frequently impossible to collect data in the form in which we wish to conceptualize and model it. A considerable variety of approaches to spatial interpolation exist and, within these, different distance-weighting and other parameters may be set. There is therefore a need for great care in the selection of an appropriate interpolator for any specific application. Such considerations as locational accuracy and distribution of the input data, output accuracy required, time and cost constraints, and prior knowledge of the desired output characteristics, will be important in making such a decision. The use of any interpolator involves the imposition of a hypothesis about the way in which the phenomena behave over space. Lam (1983) and Morrison (1974) stress the need to be very clear about the assumptions inherent in the model employed and to consider whether they are truly appropriate for the variable to be interpolated. Similarly, Goodchild *et al.* (1992) note that the data model will determine the range of processes and analyses that can be undertaken. Any interpolation by drawing boundaries, such as the Dirichlet tessellation, implicitly assumes that all significant change occurs at those boundaries and that each polygon is internally homogeneous. This is not compatible with ideas of continuously varying surface values. In contrast, a model of smooth continuous variation may be inappropriate for discrete phenomena (although whether these can truly be considered as discrete may depend on the scale of analysis). A technique for population-surface generation has been described, which allows the construction of an altitude matrix of population density from widely available zonal data.

The results of these data transformations may be stored as new coverages within the spatial database, displayed on the computer screen or plotter, or exported to other systems. Frequently, the information required is tabular or textual rather than graphic. The following chapter considers aspects of this final transformation stage, data output, in more detail.

8

DATA OUTPUT AND DISPLAY

OVERVIEW

The final data-transformation stage (T_4) identified in Chapter 4 is that of data output. The most obvious form of output from GIS is the map, and this has tended to distort the popular image of what such systems are capable of, reducing GIS to little more than an automated means of producing conventional maps. In many applications, the most important output from GIS analyses will be tabular and textual data, which may be transferred directly into other information systems without appearing on screen or paper at all. Examples of this type of operation include the data-manipulation sequences illustrated in Figures 7.2 and 7.5, where the final output is a new information product.

In this chapter we shall consider the various forms of output available and their implications for the translation of the data model in the machine into useful 'information' for the end user. As with many of the other operations already reviewed, much of the technology involved in data output has been developed in the fields of computer-assisted cartography (CAC) and image processing (IP), both in the specialized hardware used for data output, and in the conventions adopted for display and presentation of information. The display of spatial information in visual form is especially important because of its power as a medium for communication. The sophistication and quality of all the prior operations will be lost if the user is not able to accurately interpret and understand the information output.

Data output may be divided into two forms: display, which involves presentation of information to the system user in some form, and transfer, which involves transmission of information into other computer systems for further analysis. Display in graphical form concerns a whole range of issues about the ways in which we *visualize* complex data, and there is much to be learned both from the principles of traditional cartographic design and from recent developments in 'visualization in scientific computing' (ViSC). The transfer of data between different computer systems prompts

140

us to consider data standards and also some of the legal implications of passing data between organizations. Finally, we shall return to the relationship between GIS and cartography in order to consider 'the power of GIS', in the spirit of D. Wood's (1993) essay, *The Power of Maps*. Maps are not objective representations of reality, free from value judgements and unaffected by the relationships of power within a society. Instead, they are selective and value-laden documents which embody the interests and values of particular groups at particular times. GIS are increasingly assuming the roles previously held by traditional maps, and we must therefore recognize that the transformation processes we have been describing are not only subject to technical errors but are themselves affected in these same ways. It seems most appropriate to consider these issues under the heading of 'output', because it is the output from GIS (often, but not exclusively in map form) which is usually employed as a means of communication to others.

DATA DISPLAY

Data display is the final stage in the whole GIS process, and is concerned with the communication of essentially geographic information to the user. The most powerful medium for this communication is the graphic image, usually in the form of maps. A variety of equipment is available for the production of images, and has greatly enhanced the possibilities for GIS. The influence of display hardware capabilities on the form of GIS was particularly strong in the early years, as described in Chapter 2. Perhaps the single most important development was the design of visual display units (VDUs) in the 1960s which allowed the production of ephemeral graphics. These have developed a very long way to permit the animated graphics and graphical user interfaces with which we are familiar today. Major advances in visualization and interaction are still being influenced by the cost and speed of graphics and processing power (Wood and Brodlie, 1994). Early mapping programs such as SYMAP were initially restricted to hardcopy output on text line printers, using different text characters to produce different shades on the map (Figure 2.2). The NORMAP program (Nordbeck and Rystedt, 1972) was among the first to use pen plotters for the construction of lines, and led to much higher quality output maps. The basic limitation of the line printer (a raster display device) was one of resolution and the rectangular shape of the pixels. Modern raster-mode plotters (dot-matrix, ink-jet, laser) offer cartographic-quality resolution, and the choice of hardcopy output devices for GIS is now considerable (Fox, 1990).

Pen plotters are vector devices which replicate mechanically the construction of maps by hand. They are either of flatbed or drum type, and produce images by moving a pen across the surface of a sheet of paper

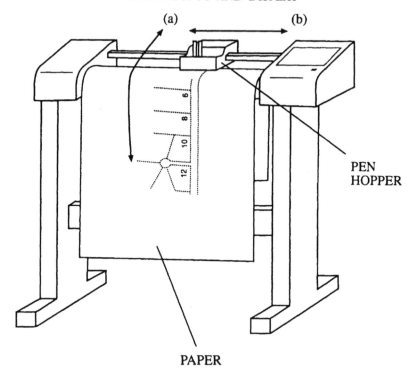

(a)   (b)

PEN
HOPPER

PAPER

Figure 8.1 A drum plotter

(flatbed) or by a combination of pen and paper movements (drum). Figure 8.1 illustrates a typical drum plotter: drawing in the x direction is achieved by moving the paper ((a) in the figure), and in the y direction by moving the pens (b). Line drawing is of high quality, but area filling is slow. The most appropriate output devices for any specific GIS installation will depend to a large extent on the nature of the applications involved. Most census mapping systems have tended to use multi-pen plotters or cheap dot-matrix methods to produce output, whereas the higher-quality devices have been limited to installations with a need for higher quality carto-graphic output relating to the physical environment.

Of equal importance to the hardware technology used is the quality of the user interface. Medyckyj-Scott (1994) stresses the fundamental impor-tance of interaction in spatial visualization. This involves everything from the design of the operating-system interface, query language and menu systems, to the use of class intervals, colour and cartographic conventions when displaying geographic data. Much user interaction with a spatial database will be in the form of specific queries and their answers. An important aspect of data output quality will be the transparency of this user interface. The majority of GIS users will have some knowledge of their data and its characteristics, but will be unaware of the collection and input of data from other sources and will not be computer scientists able to conceptualize the operations taking place within the system. For these reasons, there has been considerable interest in the development of query languages which are as close as possible to the user's natural language. Where there are explicitly geographic elements to a query, this is complicated by the need to specify location by pointing with a mouse, for example. A standard which has emerged among the database management systems (DBMS) commonly used in GIS is 'structured query language' (SQL), and an extension of SQL using a graphic interface is described by Goh (1989). Many commercial software systems are 'customized' for major users, often providing a windows-based system of menus and commands which are tailored to the user's application needs. Menu-based systems are often constructed by the software supplier on top of a more comprehensive command-level interface to the software. In this way, a number of different interfaces may be built up to suit the needs and expertise of different types of user.

Despite the wide variety of data output options available in GIS, the visual map remains the most basic and powerful means of conveying spatial information to the user. The data resulting from the manipulation operations specified will be transformed into a graphic representation on the screen or plotter, and finally interpreted by the human operator. The ability of the map reader to correctly interpret and understand the information contained in the image is therefore a key stage in the process. Much attention has been given to the acts of map reading and interpretation in conventional cartography, resulting in a number of widely adopted principles and conventions for good statistical map design (see for example Dent, 1985). The production of maps by computer overcomes many of the technical constraints of manual map production, with the potential for faster production, greater precision and range of colours, and direct computation of map parameters from the mapped data. However, the availability of the technology has not always resulted in high-quality map design, and output graphics have often demonstrated the capabilities of the hardware rather than the more theoretically desirable properties of good cartography.

In the rapid, technology-led evolution of GIS, insufficient attention has been paid to these issues, and conventional conceptual approaches to GIS have not focused attention adequately on the consideration of this final data transformation. There is a need for very careful use of display techniques in order to offer the best possible communication of information, an issue complicated by the fact that each output map is a unique product, created by the system user. A considerable degree of intelligence is therefore required at the time of software design to ensure that the user does not construct meaningless or misleading cartographic representations of the data. This will be particularly relevant to the representation and display of socioeconomic phenomena with their unique distributional characteristics, often continuously varying in both locational and attribute dimensions.

The display and interpretation of geographic data is prone to a number of uniquely geographic issues concerning design and interpretation, in addition to those shared with other forms of statistical graphics. It is probably true to say that the general principles for 'envisioning information' (Tufte, 1990) all apply to GIS output, but it is also necessary for us to consider a couple of particular problems which frequently arise when seeking to produce map representations of socioeconomic data in areal form. These are the ecological fallacy and the modifiable areal unit problem. These issues apply to data of all types which relate to individual locations but are aggregated areally during data collection (T_1) or input (T_2).

The ecological fallacy

The ecological fallacy (Blalock, 1964) lies in the fact that there are many possible aggregation strategies for a set of individual data. Relationships observed at a particular level of aggregation do not necessarily hold for the individual observations. For example, a high positive correlation between unemployment rates and immigrants in a given zone does not necessarily mean that the immigrants are unemployed. It is conceivable that at the individual level there may be a zero or very weak relationship between the two variables. In addition, the shift need not be from one distinct type of unit to another (e.g. individual persons to census zones), but may merely be a change of scale (e.g. ward to county). Although the suggestion is not necessarily that a genuine causal relationship exists at one scale and not at another, the problem is that, as the aggregation scheme changes in an arbitrary fashion, so the effect of other influential but unknown factors is varied. In larger aggregations, probability theory suggests that these unknown factors are more likely to cancel each other out, leading to more stable observations, quite apart from genuine relationships between the variables under study. The severity of the ecological fallacy therefore depends on the exact nature of the aggregation being studied, an issue which leads us to the modifiable areal unit problem. This is illustrated by Figure 8.2, in which the mean household size in a census zone is four. This aggregate value is actually found in only two of the individual households, and is totally unrepresentative of the large value in the centre. Apparent relationships between mean household size and any other variable averaged over the whole zone (e.g. levels of car ownership) are unlikely to be true of all individual households.

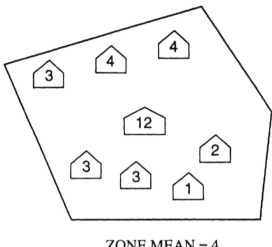

ZONE MEAN = 4

Figure 8.2 Illustrating the ecological fallacy

The modifiable areal unit problem

The related but specifically spatial problem has become known as the 'modifiable areal unit problem' (MAUP) (Openshaw and Taylor, 1981; Openshaw, 1984; Fotheringham and Wong, 1991). The key difficulty here is that there are a very large number of possible areal units which may be defined, even with the imposition of certain size and contiguity constraints, and none of these has intrinsic meaning in relation to the underlying distribution of population, hence, they are 'modifiable'. One solution, which has been implicit in much geographic work is to assume that the problem does not exist. This is certainly not acceptable in the context of GIS where the representation model is not merely an analog map of the real world, but is a dynamic digital structure which may be used as input to a variety of sophisticated analyses. The MAUP actually comprises two distinct but closely related problems:

1 The *scale problem* basically focuses on the question of how many zones should be used, i.e. what is the level of aggregation.
2 The *aggregation problem* concerns the decision as to which zoning scheme should be chosen at a given level of aggregation.

As with the ecological fallacy, once the data transformation has taken place, there is no way in which the characteristics of individuals (in this case, their locations) can be retrieved from the data. The aggregation of point records to areas cannot be undone by algorithmic manipulation. The result of these difficulties is that any apparent pattern in mapped areal data may be as

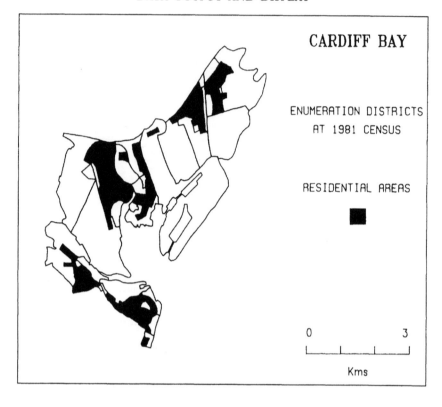

Figure 8.3 Cardiff Bay: illustrating the problems of aggregate area-based data

much the result of the zoning system chosen for the data as of the underlying distribution of the mapped phenomenon itself. Figure 8.3 is an illustration of the types of situation which may be encountered as a result of these difficulties. The 1981 UK Census enumeration districts and residential areas are shown for the Cardiff Bay region of South Wales. Some of the census EDs contain large areas of open water and industrial land, and any area-based mapping technique (such as that used to produce Figures 2.4 and Plate 1) will give a grossly misleading representation of the distribution of the population and its characteristics. The ED boundaries are unrelated to the underlying distribution and the degree of distortion introduced is hard to measure and control. The MAUP concerns both the internal representation and the final display of areal data, and is identified by Fotheringham and Rogerson (1993) as one of the outstanding impediments to spatial analysis within GIS.

146

Visualization in GIS

Visualization in this context may be thought of as 'the use of computer technology for exploring data in visual form' and also 'the use of computer graphics for acquiring a deeper understanding of data' (Visvalingham, 1994: 18). This clearly involves the use of appropriate computer hardware and an awareness of statistical difficulties such as the ecological fallacy and the modifiable areal unit problem discussed above, but it goes much further than this. The exploration and understanding of geographic data are familiar aims to geographers, and cartography has made a major contribution in this area. However, the representation of aspects of the world as data inside the computer opens up new possibilities for visualization and interaction with data. The main thrust of new developments in this area have largely come from outside geography, in a field which has become known as 'visualization in scientific computing' (ViSC). Existing visualization software is extremely powerful, but is not particularly well suited to the kinds of data model with which we are now familiar in GIS. It is interesting to note that of all the contributors to Hearnshaw and Unwin's (1994) text *Visualization in Geographic Information Systems*, only one had used exclusively GIS software, the remainder writing their own programs or using proprietary visualization software from non-geographic applications. We may therefore expect this field to become increasingly important as new techniques for the presentation of geographic data are developed and advantage is taken of the potential for animation, sound and multimedia approaches in geographic applications. The importance of the traditional map as a visualization tool is considered by M. Wood (1994).

Cartographic representation is basically concerned with the communication of the key points of geographic distributions, enabling the map reader to perceive spatial pattern, to compare pattern between different variables, and to assess contrasts between different locations. These include answering the GIS-type queries: 'what is at location A?' and 'where is phenomenon B?'. In many cases, the most appropriate way of achieving this end will not be to reproduce all the available data in graphic form, but to present a simplification which conveys the key aspects of the distributions displayed. Where precise information is required about the objects existing at a large number of locations (e.g. the address information in the example in Figure 7.2), the most appropriate form of output will be tabular, as shown, and not the production of a point map largely obscured by textual information. Dobson (1979) notes that increased map complexity will complicate the extraction of individual features and slow the process of map interpretation. This advice is especially pertinent to GIS, where the richness of the data environment is a temptation to include too much detail in the image.

Some basic principles for the use of colour, class intervals and legend

147

information are considered here. In most applications, the conventional cartographic symbolization of geographic phenomena will be of the same spatial object class as the phenomenon itself. Dent (1985) presents a table of these conventional symbolizations, and Bertin (1981) identifies seven primitive dimensions for the construction of cartographic symbolization, these being size, value, texture, colour, orientation, shape and location in space. Most geographic data undergo some form of geometric transformation during display, due to the need for generalization to fit the mapped data into the display space.

The first type of mapped data we shall consider are area-based data, which represent the majority of contemporary socioeconomic applications. The conventional way to represent such data is as a choropleth map (areas of equal value). One important principle where colours, shades or area-filling patterns are to be used to indicate the value of some attribute characteristic is that the shading patterns used should be ordered (Bertin, 1983). Despite the use of various imaginative shading schemes, colours and patterns have no natural order, and a truly ordered series can only be achieved by varying intensity, such as a grey scale, or a basic hatching style in which the distance between the shading lines varies. Figure 8.4 shows two series of shading patterns, the first of which (a) has no natural order and the second of which (b) has a logical progression from low to high shading density. A simple obstacle to the application of this principle in many low-cost mapping systems has been the restrictions imposed by early microcomputer graphics devices limited to four or sixteen (non-ordered) screen colours. The ability to compute shading intervals directly from the data values and the increasing number of ordered colours (e.g. a grey scale) which may be produced by more sophisticated raster output devices mean that it will often be possible to produce thematic maps without fixed class intervals, a form of data display suggested by Tobler (1973). Although this may seem a most attractive way of conveying continuously varying data values, commentators note that the human

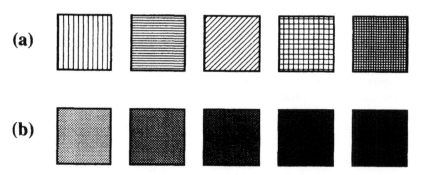

Figure 8.4 Shading series

148

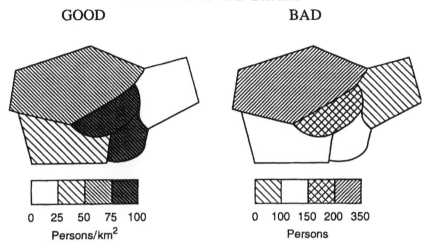

GOOD   BAD

0   25 50 75 100   0   100 150 200 350
Persons/km²   Persons

Figure 8.5 Use of shading schemes and class intervals

observer is rarely able to distinguish more than ten shading densities, and
that a series of between four and ten classes is preferable to continuously
varying shading. Simplified examples of the good and bad use of such
shading and class intervals are given in Figure 8.5. In the 'good' example,
population density (related to zonal areas) is indicated by a graduated
shading series. In the 'bad' example, absolute numbers (whose relationship
with the area of the zones is unknown) are represented by a non-ordered
shading series, giving a very different visual impression of the distribution.
The selection of class intervals is considered in some detail by Evans
(1977), and may be of one of four basic types:

1 *Exogenous:* the class intervals are selected in relation to some meaningful
threshold values external to the dataset to be mapped. This type of
classification is particularly useful if comparison is to be made between a
number of different maps of the same variable. Appropriate external
thresholds are rarely available in reality.

2 *Arbitrary:* unrelated to either the data or any external rationale. These are
generally unhelpful, and should not be used.

3 *Idiographic:* derived in some way from the data themselves. This category
includes the use of percentile groups. When used to classify zone values,
these may result in considerable differences in the areas represented by
each class. Percentile mapping should be accompanied by a frequency
histogram to allow the user to interpret the data distribution. There are
some advantages to be seen in setting the class intervals at 'natural breaks'

149

in the data range, but great difficulty in determining what exactly is a natural break.

4 *Serial*: a consistent numerical sequence, but not directly derived from the individual mapped values. This includes regular intervals on a variety of scales (arithmetic, geometric, etc), and may simply be equal subdivisions of the data range.

The most appropriate classification scheme will depend on the nature of the distribution to be mapped, but most currently available GIS software does not have any rules to govern the selection of class intervals, and either uses some default arithmetic progression or leaves the selection of intervals entirely to the user, which often results in undesirable arbitrary classification schemes. The computer-produced population maps in *People in Britain: A Census Atlas* (CRU/OPCS/GRO(S), 1980) used a system of absolute number maps and signed chi-squared maps, in which the values in each cell were classified in relation to the national average. A problem with expressing values in relation to an external average is the infinite range of possible scales over which such an average can be calculated. It is impossible to say whether classes should be relative to local, regional or national values to give the most 'meaning' to the map.

An important alternative approach to area data display, not widely applied but offering considerable advantages over many of the above, should be noted. This is 'graphical rational mapping' (Bachi, 1968). The data plane is divided into a tessellation of hexagonal cells, superimposed on the data-collection zones, and a symbol is assigned to each cell; thus large zones contain more cells, as illustrated in Figure 8.6(a). Symbols are placed in each cell, according to its estimated value, with a fixed total amount of shading divided proportionally among the cells. In zones with very low population density, the symbols in each hexagonal cell may be very small, so as to be almost invisible to the eye, whereas in zones of high density, the symbols may contain up to a hundred times heavier shading, in proportion to the actual data values (Figure 8.6(b)). This gives a far clearer indication of the distribution than any conventional choropleth map, while still using the zone boundaries as the basic georeference for the data. Bachi and Hadani (1988) describe an operational system, running on a personal computer, which uses this approach. Although essentially a method for data display, the representation of zone data in this structure is a powerful tool for more complex geographic analyses, allowing consistent interpolation and modelling of values across areal boundaries. The analysis of adjacent cells allows the identification of major clusters in the data, and other zone-independent analyses are possible, including the generation of a variety of geographic statistics for the comparison of distributions.

Another alternative approach to the display of area-based data such as population counts is the cartogram. Dorling (1994) argues cogently that

(a)

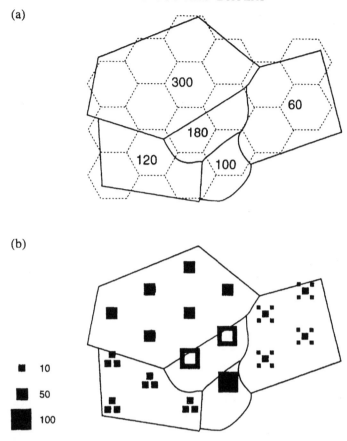

(b)

Figure 8.6 Illustrating the principles of graphical rational mapping

conventional choropleth representations of human geography give greatest weighting to those areas in which fewest people live (because zones are larger and visually dominant), and that a more appropriate technique is to devise a cartographic product which assigns visual importance to locations according to their information value. This is explicitly a cartographic solution to the problem of arbitrary zonal geographies addressed by population-surface modelling in Chapter 7 (Dorling, 1993, 1995). Plate 7 is a cartogram representation of persons who have no current occupation, and should be compared with the conventional zone-based mapping of Plate 1 and the surface-based Plate 5. In a cartogram, a particular exaggeration is deliberately chosen, in preference to the various but uncontrolled exaggerations within traditional cartographic products. Dorling illustrates a range of possibilities, including area cartograms of the UK in which area on the map is proportional to population count. The basic principle underlying

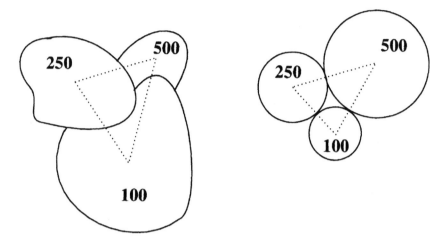

Figure 8.7 Cartogram construction

the construction of such maps is illustrated in Figure 8.7. In this example, the irregularly shaped zones are replaced with circular symbols whose area is proportional to their population counts. Such representations may be achieved by adjustment of the 'true' locations, producing either regular zonal symbols (as in Figure 8.7) or transformed area boundaries.

Attention is given to the use of graduated symbols in GIS by Rase (1987). These may be applied to either point or area data types. In the case of area data, the areal values are represented by a point symbol situated at some central point within the zone, thus creating an additional object-class transformation during data output. When dealing with point-referenced socioeconomic information such as postcode-referenced cases, it is important to ensure either that multiple cases are not obscured by overplotting or that symbols are designed to denote the presence of multiple cases. Multivariate symbols are possible, making use of varia-tions in intensity, colour, texture and orientation. These non-overlapping symbols, called 'glyphs' (Dorling, 1994), are powerful visualization tools, of which insufficient use has been made in mainstream GIS applications. Again, the production of symbols by automated means is very rapid, and there may be a tendency to condense too much information into a single multivariate symbol. This may create confusion and be an obstacle to correct interpretation of the output map. Rase suggests that as a rule of thumb no more than three variables should be indicated in a single symbol, although glyphs representing human faces or trees in which the components (facial features or branches) show the values of many different data variables create powerful multivariate maps that reward careful reading (Chernoff, 1973).

The representation of surface models in the form of output graphics is problematic, although a variety of options are available. Digital elevation models (DEMs) are often represented by drawing simulated surfaces (a two-dimensional representation of a three-dimensional object), as illustrated in Figure 8.8. These are a powerful aid to the user's visualization of the phenomena represented by the data. Different algorithms exist for the computation of these models from different surface model structures (altitude matrix, TIN, etc). A more conventional type of output is in the form of isoline (lines of equal value) or contour maps. The contour lines are usually derived directly from contours held in the database, and display involves no additional coordinate interpolation. Elevation models of this kind may also be used to derive models of slope and aspect, which provide further aids to surface interpretation. The use of raster altitude matrices offers the potential for direct output of the surface model as a classified raster image (as illustrated by the population-density surface in Plate 5). In this example, the data values in the map have been transformed into the range 0–255, which is the maximum number of shades between white and green which were available on the device on which the map was produced. For data analysis, such a map should be accompanied by a frequency histogram showing the distribution of the raw data values across the shading classes. When dealing with population-related data, it is important to ensure that unpopulated regions are preserved as a separate class, as their locations define the settlement pattern on the map. As noted above, most GIS software would not use such a classification scheme by default and

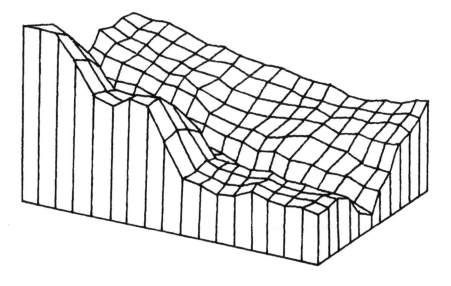

Figure 8.8 Surface representation in simulated 3-d

might involve considerable effort on the part of the user in the generation of such a display.

Approaches such as graphical rational mapping, the use of cartogram representations and multidimensional glyphs are extremely powerful tools for the visualization of socioeconomic data, but do not appear as standard display options within any of the major commercial GIS packages. GIS offer an immensely powerful environment for the exploration of alternative visualizations, yet this aspect of the data-output transformation remains relatively unexplored, except by individuals working with their own custom-written software.

DATA TRANSFER

The explosion in the collection and use of digital geographic data has led to the existence of many databases which are of interest to a range of organizations in addition to those responsible for their collection. To avoid unnecessary duplication of data collection and generation, an obvious solution, stressed in the report of the Chorley Committee (DoE, 1987), is the need for widespread data interchange between organizations with common information needs. This may either take the form of shared attribute information, extracted from a non-graphic DBMS at one site and imported to a GIS at another, or transfer of explicitly spatial information such as digital boundaries between different GIS installations. The main obstacles to free data interchange are in reality more often administrative than technical, with organizations reluctant to release their data. This may be for a variety of reasons: data collection is always an expensive exercise and there is a need to extract the maximum return on its use, sometimes there may be fears that the data is not of the high standards which the outside world may expect (!), in commercial situations there may be good reasons to prevent competitors gaining access to parts of a database, and many databases contain confidential information about individuals. While these issues persist, duplication of effort cannot be avoided. Here we shall review briefly the nature of the hardware through which data transfer can be achieved and some of the transfer standards which exist for geographic data.

Typically, an early computer installation consisted of a single mainframe processor surrounded by a collection of terminals and other peripheral equipment, with no connections to the outside world. The hardware capabilities which were available within this local installation were a very real constraint on the types of operation which were possible. The situation is now much more complex, with many installations incorporating a number of separate processors, which may need to access a central common database. The growth in the use of personal computers and workstations has increased the use of networking solutions for the full integration of information held in such systems. Consideration of the implications of distributed database architecture for GIS is given by Webster (1988). The need for data transfer between computer systems may arise within a single large organization, perhaps where databases for different applications have been accumulated on separate hardware systems. Alternatively, wider networks may be required where there is a demand for data to be shared between organizations. In the early 1990s, many computer users have become aware of the 'Internet' (Rose, 1992; Kehoe, 1994), which is not a single network but a series of connections between many computer networks worldwide. The Internet makes possible the addressing and transmission of electronic mail, searching and retrieval of remote databases, and remote interactive connections virtually anywhere. The number of computers connected to the Internet continues to double annually, with an estimated 2.2 million machines and 25 million users in January 1994 (Goodchild, 1994). This growing network provides a medium for communication, a source of information about geographic software and data, and free access to much software and data.

Until very recently, software suppliers frequently chose to work with equipment supplied by a single manufacturer, particularly for 'turnkey' systems. This tendency of computer installations to adopt hardware from a limited number of suppliers has led to manufacturer-dependence of many networking solutions (Petrie, 1989). Modern networks thus have to bridge the gaps between the standards adopted by a range of system suppliers, and the results may be very complex and expensive. Large networks may employ a dedicated computer for management of the system, and separate processors with large filestores to act as fileservers to remote processors are common. At the highest level of networking, direct communication between databases is possible, allowing (for example) GIS software running on one processor to retrieve information from a remote database while answering a query, in a way which is completely invisible to the user.

Data standards

Standards for the transfer of digital geographic information can exist at a number of levels, each of which has important implications for the use of datasets in contexts other than that for which they were originally constructed, an idea explored further by Guptill (1991). The highest level is that of the definition of the geographic objects represented by the data. Land (1989) highlights the disparities which often exist between the definitions applied by different users to the same spatial features. Examples are the classification schemes adopted for land use and the boundary definitions used for land parcels. In the context of socioeconomic information, this issue is related to the discussion of the most appropriate object class for population-related phenomena, introduced in Chapter 4. The

155

definition of geographic objects is frequently scale- and application-dependent, making the adoption of a standard set of definitions very difficult. One possible solution is the use of dictionaries of spatial data, several of which have been drawn up by agencies concerned with digital data standards such as the Ordnance Survey National Transfer Format (NTF) glossary of terms.

Another level at which a range of standards have been proposed is the level of data structures. Such standards should be transportable across a range of hardware and software systems, and the reformatting/transfer process should be as efficient as possible due to the very large volumes of data involved. As with the technology used for transmission of data across networks, the approaches adopted by major software producers have tended to be highly influential, especially in the early years of system development. Subsequent efforts in the CAC and GIS industries have been directed at reconciling these individual specifications into widely applicable standards (Clarke, 1990). Many commercial software systems use their own formats for internal data storage, but include reformatters, allowing geographic information to be imported and exported in a variety of the more common structures. A basic feature of data standards should be that they facilitate data integration from various sources in a way which is independent of the systems used and which incorporates all the information necessary for data interpretation (Cassettari, 1993). Cassettari provides a table of major international transfer standards which includes no less than thirty-two entries, demonstrating the complexity of arriving at a small number of definitive formal standards. Some standards have been widely adopted because of their use for important datasets, such as the use of the DIME and TIGER data structures (Cooke, 1989) developed for US Census of Population data. At this lowest level, standards concern the rules for the content and layout of data files which can be transferred between different software and hardware systems.

Important data formats developed by the United States Geological Survey (USGS) include standards for digital elevation models (DEM) and digital line graph (DLG) data. DLG is a vector data standard, developed for particular series of USGS maps, and involves topologically structured vector data with associated adjacency and connectivity information. The standard uses a coordinate system local to each map sheet, but includes header information with the necessary parameters for conversion into UTM (Universal Transverse Mercator projection). Experiments in the UK have led to the establishment of a National Transfer Format (NTF), which was first released in 1987 and has since been updated. The standard has five levels, defining file structures relating to different degrees of data complexity. NTF represents the work of various agencies in the survey industry, with major input by the Ordnance Survey and more recently the Association for Geographic Information (AGI) (Rhind, 1987; Cassettari,

1993). This to some extent supersedes various earlier attempts at national standards.

An example of the need for common data standards regarding socio-economic data may be seen in the development of a standard for address referencing in the UK. In June 1994 the British Standards Institute (BSI) launched BS 7666 parts two and three, which specify a standard format for address and property referencing (Cushnie, 1994). This standard for spatial datasets has been developed in three parts. Part one was published in 1993 and provided the specification for a street gazetteer, which is already being implemented by local government. Part two provides the basis for a national land and property gazetteer, which would be an index for all Basic Land and Property Units (BLPUs), where these are contiguous areas of land under uniform ownership rights. The definition of the fundamental unit for such a system has in fact proved very difficult, due to the different purposes for which organizations define land and property (Pugh, 1992; Pearman, 1993). Each BLPU is given a single grid reference and a Unique Property Reference Number (UPRN) which link it to a Land and Property Identifier (LPI). The LPI includes an addressable object name, as defined in part three of the standard, and must point to a single BLPU, although a BLPU may have more than one LPI. The third part of BS 7666 specifies a standard address structure. Primary addressable objects include BLPUs and uniquely identifiable buildings. In addition, secondary addressable objects may be defined either by subdivision or by reference to primary objects. From this brief example it may be seen that agreement over definitions for spatial data involves coordination of the interests and working practices of many different organizations, and is far from straightforward: implications of each level from the conceptual definition of objects to the arrangement of records within data files are pertinent here. At present, BS 7666 is not obligatory, but the success of such developments is to a certain extent self-reinforcing, in that as the major data-holders begin to build links to a new standard system the incentive for other organizations to adopt the same standards increases strongly.

Legal aspects

The legal issues surrounding the use and development of GIS are complex, and focus particularly on the proliferation of digital geographic data. Of particular importance are ownership and liability, which determine who may access data and who is responsible when things go wrong (Cassettari, 1993). For most users, geographic data represent information which is important to the working of their organization, yet information is a difficult commodity to quantify and regulate: it cannot easily be mea-sured, it is not diminished by use, and it is of different value to different users. In addition, the handling of digital information is (in legal terms) relatively new and evolving rapidly, and there is thus little past experience on which to base new legislation. However, values, procedures and rules relating to information have been developed over a long period, and these are being adapted to deal with the new issues posed here. These concepts are dealt with in some detail by Epstein (1991).

Maffini (1990) notes the importance of governments as the creators of much digital geographic data at public expense, but notes that policies regarding data access rights and cost recovery vary widely. In many countries there is no clear legislation concerning rights of access, and, among those countries with clear policy, the principles adopted differ widely. Most countries have legislation which protects individual privacy and therefore limits the holding and use of information which relates to identifiable individuals. In the UK, there are complex and restrictive rules governing the copying and use of state-collected cartographic data (paper or digital), and copying or reproduction may only be performed with the permission of Ordnance Survey. In the US, freedom-of-information legislation ensures that there is little restriction on the copying of paper or digital products (Dansby, 1994). The implications of these differences in approach have already been noted in the resulting differences in access and cost of census information in the two countries. Rhind (1994) argues strongly in favour of the cost-recovery principle which is being implemented in the UK, and notes that for any government the provision of spatial data is unlikely to be seen as a single policy issue (as it is by the GIS industry), but will in reality be influenced by a 'cocktail' of policies, developed for many reasons, but each having various implications for the spatial data handling.

The power of GIS for the integration of diverse geographic data means that new 'value-added' datasets are frequently produced by the combination of one or more source layers. The 'ownership' of such a hybrid dataset may be very hard to determine, and has been the subject of considerable controversy in individual cases. In principle, such disputes would best be avoided by the use of appropriate contracts concerning rights to use any data being purchased. This issue is also particularly important when data are found to be inaccurate or inappropriately used. If one of the parties is clearly in breach of a contractual obligation to provide software or data to a given standard or specification, it is relatively simple to apportion liability. In many cases, however, contracts do not exist or do not address issues of data quality, and in cases of dispute it is necessary to determine whether one of the parties has been negligent. Negligence occurs where an individual fails to take reasonable care which may be expected in a given situation, and some damage to a third party results. Clearly, there is a need to determine 'reasonable care' in the context of geographic data and for data providers to carefully anticipate potential uses of their data. Epstein (1991) illustrates a number of liability scenarios in which data providers in the US have been found (fully or in part) liable for events resulting from errors and omissions in represented locations, inappropriate representations on the use of mapped data in ways not originally intended.

We may conclude that although national governments have taken a variety of different approaches to information legislation, there is widespread dissatisfaction with the existing legal provisions which cover the trade in and use of digital geographic data. Any situation of uncertainty, for example, in which data suppliers fear that they might be held responsible for accuracy in applications which were not originally envisaged, works against general calls for greater freedom of access and sharing of information. As the market for GIS and digital data continues to expand, both the number and complexity of legal cases arising from (mis)use of such systems will grow, and this area is currently poorly understood by GIS users. As with data standards, there are considerable difficulties in achieving consensus and change in an international industry which involves so many different interest groups who have already invested heavily in specific approaches.

'THE POWER OF GIS'

We have now reviewed the various transformations by which data are processed within a GIS and have seen the central importance of a geographic model of the world. As already noted, this is not an objective process, free from error and interpretation, but is subject to uncertainty and operator intervention at almost every stage. Before concluding this discussion of the output transformation, it is appropriate to consider briefly the purposes to which such systems may be put.

Writers about cartography have been in no doubt regarding the 'power of maps' (D. Wood, 1993). In many cases, particularly in the socioeconomic realm, geographic pattern is not directly observable by eye: pattern only emerges when data are collected and mapped. The same is true for some physical characteristics such as soil type or underground pipes: we are only conscious of variation over space once we have seen the mapped phenomena on paper or computer screen. Thus when we consider the urban environment in Figure 1.1 we can see the tall building and the bus, but we are unable to *see* family breakdown or poor access to urban facilities. The mapped image of these social conditions is thus in a sense defining reality, not simply representing it. Certain areas will only receive government aid if the geographic data suggest that they fall below some threshold; therefore the decisions about which data are to be collected and how they are modelled are crucial to the actual allocation of resources. The modelled representation is actually an object of power, and the map author is thus powerful, able to define certain aspects of reality according to their purposes. These purposes will often be corporate or organizational rather than personal, but are likely to represent the interests of certain groups in society at the expense of others. The implication is not that all map makers have malicious intentions, but that the map is a powerful document and always carries values and judgements by its selection of features for inclusion, projection and symbolization. For centuries, maps have been used to impose boundaries and raise taxes, to demonstrate geopolitical claims, to achieve votes by gerrymandering, and to convey propaganda messages (Monmonier, 1991, in the aptly named *How to Lie with Maps*). There is a well-rehearsed argument in geography that as quantitative methods take only existing data, and are driven by those interest groups which are already powerful, they can only serve to reinforce the status quo: no scientific approach is in fact neutral and value-free (T. Unwin, 1992). However used, the map cannot challenge issues which it does not address. We must recognize, with D. Wood (1993: 22), that 'every map has an author, a subject, a theme'. Harley (1989), a particularly influential commentator in this field, suggests that the map is a value-laden text which may be 'deconstructed', revealing the interests which it serves.

It is necessary therefore that we should consider how these insights apply to GIS technology, which, as we have already seen in Chapter 4, creates significant extensions to the traditional cartographic process and is centred around a digital rather than an analog model of the world. One of the first things to recognize is that the illusion of scientific objectivity, already clearly seen with professional cartography, is reinforced by the use of computer technology. If anything, the GIS offers more scope for the inclusion of error and the manipulation of the mapped message to convey particular messages, but the digital medium inevitably carries more authority than the paper. This is particularly important because where mapped images are involved in conflict, it is usually conflict over the control of space in some way. While maps have often been used by influential actors, as in 'red-lining' (informally designating certain neighbourhoods as unsuitable for mortgage lending) by estate agents, the integration of diverse geographic datasets afforded by GIS permits many new forms of power to emerge. The use of geodemographic indicators for credit-scoring of consumers on the basis of their area of residence has led to some high-profile legal cases but remains an obvious application for such data. The existence of increasingly high-resolution geographic databases means that the time is drawing closer when many organizations will be able to infer with considerable precision certain household and individual characteristics appropriate to their own business interests. The potential for geographic information to effectively bypass existing constraints on the holding of personal data is complex and inadequately covered by existing legislation.

An argument constructed with a GIS, with evidence from the results of complex spatial analyses, will be harder to refute than one based on a single paper map. Monmonier (1991) clearly illustrates how maps are often redrawn many times during a battle for planning permission between developers and a local community. This is possible because the map may be manipulated to present different arguments – demonstrating the positive and negative aspects of the proposed development. Such an exchange is also possible because any group with the appropriate skills can create a map. The GIS is an even more refined tool for adjusting the message conveyed by the map (Egbert and Slocum, 1992), but it is also inaccessible to many groups. In an environment in which data costs and system investment are rapidly emerging as the real entry barriers to GIS use, these tools will almost always be in the hands of the already-powerful, typically government and commerce. There has (as yet) been little evidence of 'radical GIS', constructed to present alternative views of space, and the practicalities of such a system would be extremely challenging. Sui (1994) characterizes GIS as a throwback to the positivist mode of scientific enquiry common in the human geography of the 1960s, but stresses that the gulf between GIS practitioners and critics is not unbridgeable. The GIS community is already showing some concern for the ethics of system application and interest in fuzzy logic and concepts of relative space, but these discussions tend not to be couched in the same terminology as many of the contemporary critiques.

We can conclude, with Innes and Simpson (1993: 230), that GIS is a 'socially constructed technology' that does not simply exist in terms of hardware, software and data, but incorporates institutional goals and practices. The more successful the installation in operational terms, the closer these relationships are likely to be. This knowledge is not made explicit throughout our examination of the workings of GIS, but cannot be ignored. If, in the words of D. Wood (1993: 188), 'maps are heavy responsibilities' we may conclude that GIS are heavier.

SUMMARY

This discussion of data output has considered both the final display of geographic information to the GIS user and the intermediate transfer of data between information systems. These output operations form the final stage in the transformation-based model of GIS operation. The quality of display determines the ease of interpretation of the output information and has enormous potential for either educating or misleading the user. A wide range of very high-quality vector and raster output devices is available, and the cost of hardware may be expected to continue to fall. However, the ease of creation of complex graphics, and the data-rich GIS environment tend to lead to the creation of badly designed output graphics with too much detail and too little attention to the actual meaning of the information content. The use of expert-systems interfaces to GIS (Webster, 1990) may offer hope of increased control over the output generated in response to user enquiries. Important new developments are occurring in the field of visualization in scientific computing (ViSC), but at present these are not well suited to the handling of geographic data. Innovative visualization tools such as cartograms, glyphs and graphical rational mapping have tended to remain outside the mainstream software systems, which concentrate on reproducing that which is already familiar.

Many of the issues relating to data transfer between organizations would appear to be 'common-sense' solutions, for example to avoid the unnecessary duplication of digital data or the maintenance of related records in incompatible formats. Unfortunately, each individual GIS supplier or user is faced with pressures and circumstances which often prevent their adopting these most desirable solutions. Major de facto standards have emerged through their use by best-selling software systems or for important national datasets. Increasingly, major systems provide tools for the conversion of data to other formats, which permits relatively free interchange of geographic data. The growth of networking and the recent emergence of the Internet are also serving to make routine direct connection and transfer between remote systems. Although technological barriers to data sharing are being removed, legal uncertainty about ownership and liability persist, and this uncertainty tends to work against the data sharing which is generally seen as desirable. Our overall conclusion regarding data transfer is that the power of existing technology and geographic databases offer an enormously rich source for information communication, but their use is currently hindered by organizational issues. It is in these areas that there is a need for clear direction and a willingness to accept, wherever possible, the wider needs of the community rather than those of the individual agency.

Finally, we have noted that data in GIS are not value-free. At all stages, from the initial selection of phenomena to be included in the geographic database through to the design and presentation of output graphics, these are systems which can convey powerful messages, and which always embody the values and decisions of those involved in their construction, just as with conventional cartographic products. There is a tendency to ascribe an undeserved scientific objectivity to data held within information systems, and GIS are no exception here. This frequently has the effect of preserving the existing balance of interest groups and seats of power in society. We should recognize that in order to address many important questions concerning commercial or government policy, it will be necessary to challenge the most basic assumptions on which such systems have been constructed, and to seek complementary explanations in those features of the world which are not, or cannot be, represented by our data models. In the following chapter we shall focus more specifically on the use of GIS in the representation of socioeconomic phenomena, and will return to this theme of the value-laden nature of GIS activity.

9

TOWARDS A SOCIOECONOMIC GIS

OVERVIEW

In Chapters 5 to 8 we have examined the four transformation stages which make up our conceptual model of GIS operation. The powerful influences of physical-environment applications and hardware capabilities on the current form of GIS have been noted, and attention has been drawn to the particular difficulties faced when applying this technology to socio-economic data. Until now, the focus of the discussion has been on under-standing GIS, using population data as examples where appropriate. Finally, it is necessary to turn in more detail to our main interest in the use of GIS for the modelling and analysis of the socioeconomic environment. Plane and Rogerson (1994) identify the recent combination of population infor-mation and selected GIS functionality into what they call 'demographic information systems'. Their approach stresses the importance of the geographic dimensions of population analysis and it is argued that the potential for the application of GIS techniques in this context have not been fully realized in the implementations already cited. The technology reviewed offers a number of possible routes for the implementation of a 'socioeconomic GIS'.

An important factor in the development of techniques to handle this kind of geographic information has been the growth in geodemography, reviewed in Beaumont (1991) and Brown (1991). Geodemographic tech-niques have been primarily concerned with the classification of localities into neighbourhood types (an idea introduced in Chapter 8). Early classi-fication schemes were relatively crude analyses of census data, but there is increasing demand for specialized area classifications which will aid in retail-site analysis, credit rating and target marketing. Many of the currently available products are customized according to the needs of a particular application, and incorporate both census data and other more specialized datasets from within the organization using the indicators. Meaningful classification schemes rely on the availability of suitably georeferenced socioeconomic data. Despite the subjective nature of geodemographic classification schemes, there is clearly the greatest chance of commercial success if information can be provided in a spatially disaggregate form. The need to define the catchment area of a particular store or to delimit a residential neighbourhood with certain household characteristics requires a data model which incorporates all the important geographic aspects of the population and its characteristics. These same types of information will be invaluable to the planner concerned with health care or education, for example, and the potential uses of a high-quality model of the population should not be underestimated.

As noted in Chapter 1, by 'population data' we are here referring to data originally relating to individual members of a population which is scattered across geographic space. These data may relate to individual persons enumerated at the time of a census or may represent specific items of information about individuals with some characteristic of interest such as in a mortality register or the results of a survey. Flowerdew and Openshaw (1987) consider the problems of transferring data from one set of areal units to another incompatible set. The whole of their discussion may be set within the framework outlined in Chapter 4. They point out (p. 3) that 'physical data exist or are captured in a very different form from socio-economic information. Moreover, the objects for which data may be collected are fundamentally different.' It is the nature of these differences and the most appropriate way of dealing with them which form the focus of the following discussion. The most basic item of information is that available from a complete enumeration of the population; a fundamental question is 'what is the true spatial-object class of a population in geographic space?'. There are three possible object classes for which such data may be presented, namely point, area and surface. These are reflected in the strategies available for database construction.

The following section considers the possible approaches to the construction of GIS tailored specifically to population-related data. Each of the options are then examined in more detail, including reference to existing implementations, and consideration of their scope as a basis for socio-economic GIS development. Finally, some tentative judgements are made as to the 'best' solution and likely future developments.

INFORMATION-SYSTEMS APPROACHES

It is apparent from our review of data-storage structures (Chapter 6) and data-manipulation operations (Chapter 8) that representation of real geographic phenomena in a data structure involves the imposition of some kind of model of what that reality is like. Each of these models makes assumptions about the way in which reality is structured. A GIS geared towards the representation of the socioeconomic world must adopt one of three approaches, which are considered in the following sections. The first of these is an individual-level approach, in which data are held relating to every individual person in the population. Databases of this kind tend not to be strictly geographic, due to the difficulty of assigning unique spatial references to individual members of a population, but we may increasingly expect to find household-level georeferencing in operation. In many datasets which are concerned with the management of the urban environment in terms of property or taxation, this will be the natural level of georeferencing. The second, and currently the most common approach, is areal aggregation which includes conventional choropleth census mapping. The inherent assumption of any such method is that geographic space is divided into internally homogeneous zones, with all change occurring across zone boundaries. Census-type data are generally available already in this form. Again, for some applications concerned with the built environment involving, for example, blocks defined by street intersections, this may be the obvious division of geographic space. The third option begins with the assumption that some socioeconomic phenomena of interest to the geographer are essentially continuous over space, and attempts to reconstruct this continuity, albeit using discrete data structures as an approximation. These three approaches represent very different ways

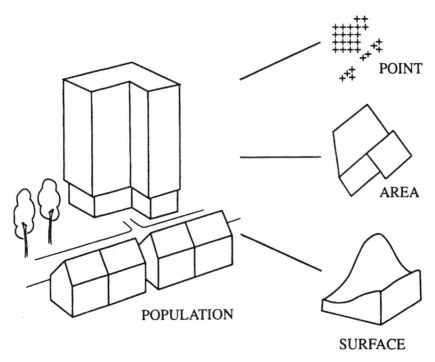

POINT

AREA

POPULATION

SURFACE

Figure 9.1 Different conceptualizations of population-related phenomena

165

of thinking about population-related phenomena (Figure 9.1), and affect the form of all subsequent operations on the data. The issue of basic spatial units (BSUs) for use in GIS is considered by Openshaw (1990), who identifies some of the difficulties involved in bringing about change in the standards used for georeferencing.

Of the various applications of GIS technology to population data used as examples in the previous chapters, none provides an ideal single model of the socioeconomic environment. Population-related data can only ever be indirectly georeferenced, and true geographic patterns in the data are obscured by the idiosyncrasies of the zoning scheme or postal system through which they have been assigned a geographic location. Some data concerning the broader socioeconomic environment are not strictly popu- lation-based, but relate to properties or neighbourhoods, in which case it is important to recognize that different basic spatial objects are involved. In Chapter 4 it was argued that it is important that a data model should be of the same spatial-object class as the user's concept of the phenomenon which it represents. The economist interested in regional-level unemploy- ment structure will find individual-level data of little help. By contrast, regional aggregations tell us nothing about local patterns. In order to be of use, a database must therefore be appropriate for the desired analysis, in terms of both scale and object class. As more geographically referenced data become available, and pressure for the processing of socioeconomic information in this form increases, it is important that the best possible approach to representation be adopted. Improving the quality of data models is the only way to avoid fundamentally flawed analyses and to facilitate truly integrated GIS, able to model both the human population and its physical environment.

THE INDIVIDUAL LEVEL

The addresses of all individuals represent the greatest level of detail possible (point observations), but usually these will not be available for mapping or input to GIS, as the collection operation T_1 involves aggrega- tion to some areal unit whose boundaries will not always be precisely defined. In some European countries, where population registers are maintained, limited statistical information will be held for all individuals, but this is not the case for most censuses of population. The strategies adopted by different governments for the collection and release of popula- tion-related data are discussed in more detail by Rhind (1991). Even if individual-level information is held by the national census office, confiden- tiality restrictions usually prevent its release in disaggregate form. In the 1991 UK Census of Population Small Area Statistics (SAS) data were suppressed for all EDs containing a total of less than sixteen households and fifty individuals (OPCS, 1992a), and in the published figures, data for suppressed zones were combined with those of a neighbouring zone. These thresholds have been raised since the 1981 Census due to the greater number of cross-tabulations which are available, which might make possible the inadvertent disclosure of information about identifiable individuals. The main attraction of individual-level databases is that they facilitate ad hoc aggregation, allowing the design of areal units to suit analytical requirements, including comparison over time and interpolation between incompatible zoning systems. Knowledge-based systems have been suggested for the management of census data which would contain an individual database, but only release aggregate data for user-specified zones, where these were not in conflict with confidentiality restrictions (Wrigley, 1990). Although confidentiality constraints prevent the release of official statistics at the individual level, much georeferenced information is collected and available commercially by postal addresses and postal divisions. Systems based on actual addresses face significant problems in terms of text recognition and ambiguous addresses, but the advent of national address georeferencing systems such as Pinpoint's address locations and Ordnance Survey's ADDRESS-POINT (Ordnance Survey, 1993) in the UK, suggest a future in which much more geographic detail will be attached to existing databases. The ZIP Plus 4 data in the USA also represent a significant step towards smaller geographic units (Thrall and Elshaw Thrall, 1994). To date, postal systems (described in Chapter 5) have emerged as the nearest approximation to individual-level georeferencing in the UK. This situation is not universal, as many countries have postal codes which are simply based on existing administrative geographies. Postcodes were recommended as the basic spatial unit by the Chorley Committee (DoE, 1987), and by the Korner Committee, concerned with the management of UK health-service information (Korner, 1980). An additional issue which is relevant at this level is the location chosen for the georeferencing of a mobile individual. Should data relate to the home address, the workplace, or both? Some variables will be more meaningful at one location than the other, but home addresses are generally considered to be the individual's 'location' for all purposes.

Personal data are currently held by many different types of organization, including those concerned with finance, health, utilities, vehicle registration, and taxation. The creation of a national data-registration system was briefly considered as part of a review of options for the collection of UK national population information, but was discarded in favour of the continuation of the existing system of periodic censuses (OPCS, 1993). For most organizations, the original rationale for creating their personal databases was the management of an administrative task such as record-keeping or billing. Often, these databases have been created ad hoc by the organizations themselves, and no single organization holds the definitive record of persons or addresses. Textual addresses are often the only direct georeference available on such records, but the advent of automated address-matching systems and the inclusion of such tools within some GIS software (e.g. ESRI, 1993b) makes effective record-matching possible, and thus the integration of many diverse databases. The development and implementation of data standards for addresses and household georeferencing, such as BS 7666 discussed in Chapter 8, will make the creation of directly geographically referenced personal databases much more easily attainable, although Martin *et al.* (1994) highlight the many inconsistencies which emerge when attempting to match address databases compiled by different organizations for different purposes. The success of individual-level databases is heavily dependent on the ability to ensure that they give complete coverage and on their being constantly maintained. Individual-address georeferencing is a particularly powerful tool in many contexts, for example the maintenance of property-ownership records or the integration of utility digital mapping with customer information. It is important to note, however, that the characteristics of a zone, however small, cannot simply be assigned to an individual whose address falls within that zone. The ecological fallacy discussed in Chapter 8 is directly relevant to the integration of individual-level data with any aggregate dataset.

Many new value-added digital data products will undoubtedly follow from these recent developments. Household-level georeferencing has the potential to effectively circumvent anonymity by allowing the matching of personal records for individuals living at the same address. It is doubtful that existing law on copyright and the holding of personal data provides adequate guidance for the many new situations to which this may lead. It is not surprising, therefore, that the prospect of the widespread exchange of personal data in this way causes public concern about the uses to which such data may be put. There are considerable variations between countries in the political acceptability of centralized holdings of personal information, illustrated for example by opposition in the UK to the introduction of any kind of personal identity cards, while these are widely accepted elsewhere in Europe. Another contrast would be the inclusion of a question concerning income on the US census questionnaire, while such information would be unacceptable in the UK and many other European countries.

In conclusion, it may be observed that individual-level databases are frequently encountered, but to date these have rarely been geographically structured. Thus truly geographic manipulation of such databases is rarely possible. Although a non-geographic DBMS may offer the ability to extract all customers in a particular county by searching for that county in the address field of the database, it will not facilitate any form of explicitly spatial query, and cannot support queries involving concepts such as adjacency, connectivity or spatial coincidence. New georeferenced household databases are beginning to emerge. In the future, the role of individual-address referencing will almost certainly grow, but this cannot occur until widely accepted geocodes are available for all addresses, and until a large user community has access to both detailed data and adequate processing power. Such databases have an important role to play in the functioning of certain organizations, but do not at present form an ideal framework for the construction of a general GIS for socioeconomic data, particularly as regards the need for confidentiality. In addition, it may be argued that there is a need for carefully designed new legislation to regulate activity in this field. Individual-level referencing will certainly increase for the purposes of operational management of customer lists and marketing data, but there seem to be a number of more fundamental issues preventing its use as the basis for general-purpose GIS concerned with socioeconomic data. The establishment of definitive address databases by government or commercial organizations may well be some years away in countries where extensive address registers do not already exist. In these situations, the continued use of geographically aggregated information is inevitable.

GEOGRAPHIC AGGREGATION

This is by far the most common approach to the handling of population data, as illustrated by the various examples cited in previous chapters (see, for example, pp. 41–45). The majority of contemporary socioeconomic datasets are released in this form, although frequently using different sets of areal units for aggregation. The assumption inherent in representing such phenomena by discrete areal units is that all significant change occurs at the boundaries and the data-collection zones are internally homogeneous. This is a fundamental flaw, as neither the attribute characteristics (age, wealth, health) nor the distribution of population can reasonably be expected to be uniform within any arbitrarily defined areal unit. Morphet (1993) examines the boundaries of 1991 UK Census EDs in Newcastle-upon-Tyne and evaluates the extent to which they coincide with significant 'boundaries' in the socioeconomic geography of the city. He concludes that, even for spatially clustered variables such as tenure type, there is little justification for the use of such boundaries as representations of the socioeconomic data. Although total counts and summary statistics for the attributes will be correct within each zone, this information is impossible to interpret precisely. The areal units chosen for census data are generally designed for ease of enumeration and only secondarily standardized for population and area. This causes difficulty in evaluating the errors associated with any given map as 'the process leading to error in each boundary depends on the nature of the boundary' (Goodchild and Dubuc, 1987: 167). Consequently a single census district may include large areas of agricultural, industrial and commercial land or open water which are completely unpopulated, as illustrated by Figure 8.3. Also, a distinct settlement may be split between several such areas, so that characteristic features of its population become masked in the aggregate data. These problems are compounded in any specific case by the fact that the relationship between the areal units and the underlying variable is essentially unknown. At the heart of these difficulties is the spatial-object-class transformation, which occurs at the data-collection stage, and the associated ecological fallacy and modifiable areal unit problem (MAUP), which were introduced in Chapter 8. It is important to realize the enduring nature of these difficulties. The ecological fallacy will always be present in data which have been through any transformation involving aggregation, as the original detail can never be retrieved by computation. The MAUP will always be present with areal data, regardless of how the boundaries have been derived (whether precisely digitized or computed geometrically in some way). Martin and Gascoigne (1994) consider obstacles to the analysis of change over time which are presented by changing census geographies; many of which are particularly pertinent to the use of geographic aggregations. It is not at all clear how to separate those apparent changes which are artefacts of the changed zoning system from those which represent real alterations to the structure of the population which is being represented. In the following two sections, vector- and raster-based methods for population representation are described. It should be noted that many of the underlying principles are common to both approaches.

Vector-based approaches

The simplest vector-mode presentation of census-type data is the display of data aggregated to the census zones in the boundaries of those zones themselves. The display of data in this way is basically an automated reproduction of the conventional choropleth map product, and is thus prone to all the weaknesses outlined in previous chapters. Nevertheless, the use of automated means to produce census atlases has been widespread, and there has been work on the design of automated census mapping systems, reviewed in Chapter 3. In any mapping exercise, it is desirable to use the smallest areal units for which the data are available (Rhind, 1983), due to the areal aggregation problems outlined above. Plate 1 illustrates a choropleth map constructed in this way. The display of count data for such zones is potentially highly misleading, as very small data values may contribute large areas of the image. The only meaningful way to present population data for these aggregate areas is to display the population as a density value per unit land area, and other variables either as densities per unit area or as percentages of the total population for the zone. Examples of the presentation, either by manual or automated means, of maps of this type are widespread (Barr, 1993b). It should be seen from these observations that the sole use of existing vector choropleth mapping is not a promising basis for the development of integrated socioeconomic GIS.

An alternative is to use a composite of available land-use information in an attempt to recapture some of the geography of the settlement pattern, as illustrated in Figure 9.2. This technique is known as 'dasymetric mapping', and enables the representation to move away from the original data-collection units. If digital land-use boundaries are available (e.g. from a classified RS image), these may be superimposed on the population-zone database, masking unpopulated regions, and data are presented only for those areas which are known to be populated. The difficulties here are in obtaining a suitable land-use base and the need for a polygon-overlay operation and areal interpolation in order to create the boundaries of populated zones. Once created, the digital map will be more complex than the original zone database and therefore larger and harder to

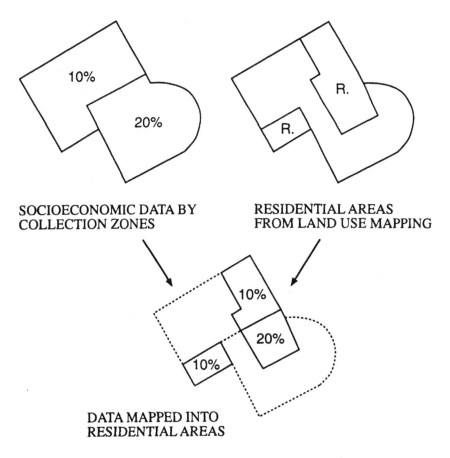

SOCIOECONOMIC DATA BY
COLLECTION ZONES

RESIDENTIAL AREAS
FROM LAND USE MAPPING

DATA MAPPED INTO
RESIDENTIAL AREAS

Figure 9.2 Population mapping using dasymetric techniques

manipulate. Langford and Unwin (1994) consider the use of such dasymetric mapping in the production of population density maps for northern Leicestershire. They note, however, that all dasymetric maps are to an extent arbitrary, according to the nature and scale of the ancillary information used. In regions classified as residential, but with variations in the population density, this approach will offer little improvement over the conventional zone-based choropleth map. In a region of scattered populations on an essentially unpopulated background, the technique will offer a considerable improvement in geographic information over choropleth techniques. This is not simply a display mechanism, as it effectively redefines the areal units to which the population values relate, and may be used to alter the geographic database.

In some applications, zone centroids have been used as the locations of point symbols to represent the characteristics of each zone, as the symbols may be expected to fall in populated areas, thus conveying some idea of the population distribution in the mapped area, although this method has generally only been applied to small-scale mapping, and is again too restrictive to provide a suitable basis for an operational GIS. A number of extensions to this approach, using multivariate symbols, are reviewed by Dorling (1994) and noted in Chapter 8.

In an attempt to produce choropleth maps at high resolution without the need for extensive digitizing, some work has focused on the use of Thiessen polygon approximations to true zones, using the boundary inter-polation technique introduced on pp. 133–135. This approach is useful in situations where zone boundaries are not available but there is some reference point for each zone whose location is known. Using the procedure described on pp. 133–135, all locations are assigned into the zone whose reference point is closest. One advantage with this type of auto-mated procedure is that the areas of zones, although only estimates, may be easily obtained by geometric calculation from the relatively simple boundary information. Where higher-level boundaries are available, the polygons may be clipped to the edges of the surrounding zones, making use of empirical data as far as possible (producing zone maps which appear as shown in Figure 9.3). Boundaries generated in this way contain far fewer points, and are thus more efficient than those digitized by hand in greater detail. Fractal enhancement has been suggested as an appropriate way of making the straight zone boundaries appear more realistic (see, for example, Batty and Longley, 1994). This type of model is best suited to applications where reference points are geometric centroids and zones are regularly shaped.

A potential application of this technique would be the construction of areal units relating to UK unit postcode locations, which currently have no defined boundaries. Using the locations of individual addresses from each postcode would allow the generation of more detailed polygon boundaries,

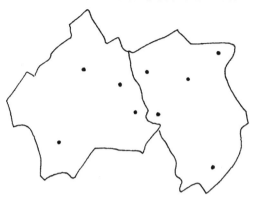

KNOWN BOUNDARIES AND SMALL-AREA CENTROIDS

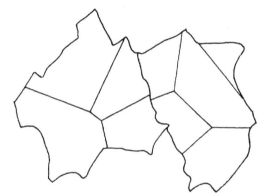

**SMALL-AREA BOUNDARIES GENERATED
USING THIESSEN POLYGONS**

Figure 9.3 Small-area boundary generation using Thiessen polygons

which could be used for the representation of data relating to each single postcode. Boyle and Dunn (1990) demonstrate the use of this technique with Pinpoint address-code data, building up unit postcode zones comprising the Thiessen polygon around each address location contained in that postcode, a method illustrated in Figure 9.4. It will still be necessary to estimate the values of all variables which are only available at higher levels of aggregation. It is suggested that this last approach, especially when applied to such high-resolution data as the PAC or ADDRESS-POINT locations, is the most promising of the areal aggregation-based methods as a basis for GIS. If widely implemented, this approach would be a powerful tool for the manipulation of socioeconomic information, although few major socioeconomic datasets are available at such a disaggregate level.

173

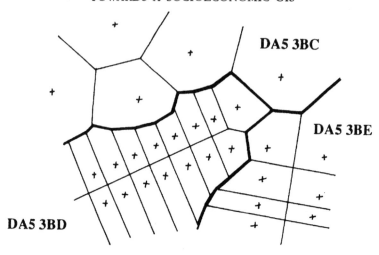

Figure 9.4 Unit postcode boundary generation

There is also considerable potential for the use of such methods in the design of future census geographies, where the lowest-level units do not currently have clearly defined boundaries. The newly available UK Ordnance Survey (GB) ADDRESS-POINT data will offer national coverage of property-level georeferences, which may be used in a variety of ways for the automated generation of higher-level boundaries such as unit postcodes and part postcode units (PPUs).

Each of these vector representations fails at some point because the geography of the settlement pattern has been lost by the transformation of point- to area-class information at the initial data-collection stage T_1. However sophisticated the technique used, the areal representation of these data will present difficulties in subsequent manipulation and interpretation because the fundamental spatial characteristics of the data are not of the same spatial object class as the real-world phenomena they represent.

Raster-based approaches

Various raster-based approaches may be adopted which are broadly equivalent to the vector methods considered above. These include the direct representation of a choropleth map in raster form by rasterizing. In this case, the underlying data model is exactly the same. This approach should not be confused with the aggregation of raw census data to grid squares, as was done with the 1971 UK Census. Data were aggregated both to EDs and to 1 km grid squares, which were simply an alternative area aggregation system (CRU/OPCS/GRO(S), 1980). A problem with grid-square aggregations is that data for many of the cells have to be suppressed for confidentiality reasons, because they contain very small numbers of people. This is of particular importance as it obscures the location of truly unpopulated regions by the addition of suppressed cells. The population distribution curve for 1971 grid squares was highly skewed, with over a third of all squares containing zero population, and a maximum in central London of 24,300. However, a great advantage of using grid squares is that they remain stable between censuses, allowing direct analysis of population change. This potential was lost when the 1981 Small Area Statistics were only made available for EDs.

Raster equivalents of the point and Thiessen models of zone data distribution are illustrated by the BBC Domesday Project. This project utilized two techniques for assigning 1981 population values to the cells of a series of rasters, ranging in size from 1 to 10 km. The changes in the apparent population distribution as the data are recombined at different cell sizes is in itself a powerful illustration of the modifiable areal unit problem. When examined carefully, the techniques used will be seen to be almost direct equivalents of the approaches already outlined. The following descriptions are derived from Rhind and Mounsey (1986), Flowerdew and Openshaw (1987) and Rhind and Openshaw (1987).

The first approach was simply to assign to each cell in the raster overlay the total value of all ED centroids falling within that cell. This is a very crude and rapid approach to the problem. If an attempt were made to create a raster image with a finer resolution, all cells except those containing ED centroids would be assigned zero population. In a reconstruction of the 1 km grid square information from the 1971 Census using this method, only about 35 per cent of the populated grid squares were assigned non-zero population values. This is clearly unsatisfactory for any attempt to construct a raster layer with a cellsize likely to reveal detailed local patterns. This approach is basically equivalent to the use of point symbols at centroid locations, in which the symbols are certain cells in a regular grid.

The second approach was rather more sophisticated, and used a Dirichlet tessellation of all centroids, assigning ED population to the Thiessen polygons and calculating the proportional contribution of each polygon to corresponding grid cells. Some areas, known a priori to be unpopulated, were then restored. The underlying assumption is one of uniform population density within EDs, with all significant change occurring at polygon boundaries, a feature of all Thiessen polygon approaches. This method gives a much more satisfactory result when applied to the 1971 data, but again no detail could be provided at a larger scale. For technical reasons, this approach was not applied to the production of Domesday data for the whole country.

These last two methods for data assignment to cells represent two alternative assumptions about the underlying distribution: the first is that all the population is present only at centroid locations, and the second is that the population is present at all points in the plane and that population density is uniform at all locations associated with a particular centroid location. In reality, the distribution of population displays characteristics which fall somewhere along the continuum between these two extreme positions. Although software containing the functionality required for each of the approaches considered here is readily available, the manipulation and analysis of areally aggregated data will always be problematic. This is essentially because geographic manipulation operations only act on values associated with the arbitrarily defined zones rather than on a model of the underlying socioeconomic phenomena themselves.

MODELLED REPRESENTATIONS

Each of the above areal aggregation approaches attempts to reconstruct some form of zonal system (either by directly digitizing collection zone boundaries or by attempting boundary generation from related data points). The aggregate vector approaches in particular, and more recently the raster methods, have been the most common strategies for the handling of socioeconomic data in cartography and GIS. As the number of commercially available digital geographic datasets (zone boundaries of various kinds, property locations, etc.) increases, the number of options which may potentially be used for the indirect georeferencing of population data will multiply accordingly, but all are fundamentally bound by the mismatch between these spatial objects and the underlying phenomena being manipulated. The remaining alternative is to attempt to model the distribution of the underlying phenomena, regardless of the collection zones. There has been increasing attention paid to a variety of different techniques which attempt to achieve digital representation of such phenomena independently of the data-collection zones, and these are reviewed by Goodchild *et al.* (1993a). Much of this renewed interest in continuous representations has arisen due to the need faced by many researchers to translate socioeconomic data from one set of areal units to another. The concept of modelling distributions is a familiar one in terms of attribute data values, as illustrated in the discussion of geodemographic techniques in Chapter 7. This idea may be extended to the locational aspects of the data. The generation of surface models from population-weighted zone centroids introduced on pp. 135–137 is an example of such modelling. The following sections address issues of conceptualization and implementation of these models.

Concept

The concept of the density and attributes of a population distribution as continuously varying phenomena has been suggested in Chapter 4. The advantage of modelling socioeconomic phenomena as surfaces is that manipulation and analyses may be performed independently of any fixed set of areal units, and modelled values are free from the confidentiality and complexity of individual-level data. When held as altitude matrices, these models are ideally suited to handling in a raster GIS environment. Surface models of such data have traditionally been represented by isopleth maps, which assume the existence of a continuous density function which may be interpolated at any point. In their review of areal interpolation techniques, Goodchild *et al.* (1993a) note that the problem of transferring socioeconomic data between different sets of areal units is best considered as one of estimating one or more underlying density surfaces. The modelled approaches discussed here are effectively tools for constructing from input zonal data general-purpose surface representations which are suited to manipulation within GIS.

Schmid and MacCannell (1955) stress that as the data for isopleth mapping are based on predefined areas, the size and shape of these areas will have profound influences on the form of the surface constructed. If the base areas are large, meaningful variations in the underlying phenomena will be masked out. The nature of the data-collection zones (which are already a transformation of the original data) will determine the usefulness of such a representation. Also, the issue of spatial scale is very important – indeed the question of the continuity of population density may be one which is scale-dependent.

At this point, some reference should be made to trend-surface analysis, as an example of an early approach to the representation of population-density as a surface. The basic aim of trend-surface analysis is to attempt to decompose each data value into a regional trend component and a local residual, by a simple modification of the regression model, using power or trigonometric series polynomials (Norcliffe, 1969). This type of construction is best suited to surfaces with a linear trend and few inflexions. Although population at the finer spatial scales is highly complex, some attempts were made to use such techniques to model population-density changes over large areas. Chorley and Haggett (1965) recall the ambiguities presented by areal data and suggest the use of trend-surface models to disentangle the regional and local variations, and the ascription of causal mechanisms to the different components. They draw the parallel between isarithmic maps describing continuous surfaces in physical geography (e.g. elevation and isobaric pressure) and the potential for constructing population surfaces: 'Population, like light, may be profitably regarded either as a series of discontinuous quanta or as a continuum. The choice is largely a matter of scale, convention and convenience . . .' (p. 48). Despite these arguments, it seems that no clear illustration of the continuity in space of population density had been given until Nordbeck and Rystedt (1970). They identify two types of spatial function: point and reference interval.

177

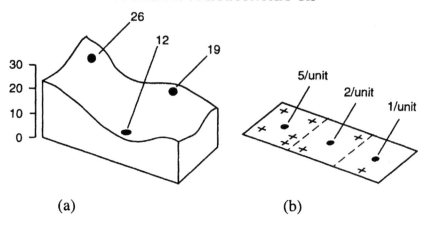

Figure 9.5 Point and reference-interval functions

Both of these may be demonstrated to vary continuously over space, and their calculation is shown in Figure 9.5.

1 Point functions (Figure 9.5(a)) are those functions of which it is possible to obtain the value $f(x, y)$ at a point location (x, y), solely by reference to that point. This assumes both continuity (variable values can be imagined to exist everywhere on the surface), and the absence of any sharp discontinuities. A common example would be land elevation above sea level, although the possibility of a cliff overhang (i.e. more than one surface value at the same location) is a violation of the strict mathematical properties of such a surface.

2 The opposite of a point function is a non-point function or, more helpfully, a reference-interval function (Figure 9.5(b)). In this case, the value of the function $f(x, y)$ depends not only on the point location, but on a spatial reference interval belonging to that point and, by definition, itself containing many points. A physical example of such a function is atmospheric pressure (force per unit area). Population density (i.e. persons per unit area) is also a function of this type.

Implementation

Unfortunately, the ideal starting point for density calculation is again a knowledge of the location of every individual's dwelling, and in reality only rough locational references such as the administrative area to which an individual belongs are available. The basic requirement is still for a set of (x, y, z) observations, consisting of a point location and a population value associated with that location. The population-density function is simply defined as the number of individuals living in a reference area located around each point. Population density may be determined at each point and interpolated for the construction of a surface model.

The early methods for population-related surface construction suffer from two major weaknesses. The first of these is a failure to preserve the correct population volume under the surface. The second is the failure to reconstruct unpopulated regions. We know from our experience of the real world that there are many locations, even within an urban system, which have no population. Such unpopulated areas would be immediately apparent in a point representation of the individual-level data, if it were possible to accurately georeference and map such a distribution, as illustrated by Gatrell (1991). These unpopulated areas may often be contiguous areas greater in size than any single zone in the zoning system, yet they are totally absent from the collected zonal data and from surface models based on these zones.

The issue of volume preservation is addressed by Tobler (1979), who suggests a pyncophylactic (volume-preserving) technique for surface construction, based on geometric zone centroids. This technique, although preserving the total population in the distribution model, assumes that the population-density function is concentric around the geometric centroid of each zone and shares the characteristic of the other models that imply the presence of population at all locations on the plane. Conventional surface-modelling techniques have been unable to preserve unpopulated regions because, however small the areal units used, there are no unpopulated zones recorded in the area data on which they are based. As a result, there is no background of zero-valued grid points for interpolation. There are two issues here: as the areal units become smaller and smaller, they become closer to the representation of individuals, but the crucial factor is less one of areal-unit size than of having zones which provide a definition of the unpopulated areas, at an appropriate scale. This has a parallel in the use of reference periods in time series analysis as aggregations of a series of discrete events. The use of irregular or large intervals will obscure the detailed form of the distribution.

The surface-generation methods described by Langford and Unwin (1994), Goodchild et al. (1993a) and Bracken and Martin (1995) overcome both of the above difficulties, and offer many attractive features for the implementation of a GIS for socioeconomic phenomena. The maintenance of socioeconomic surface models in raster form is an enormously flexible structure for data analysis. Langford and Unwin (1994) use the logic of dasymetric mapping, discussed above, in order to reassign population values from source zones into the cells of a regular output grid, using a layer of ancillary information, in this case derived from a classified satellite image. Martin and Bracken (1994) describe a series of national surface models for the UK, from which population data may be extracted across the Internet for any user-specified region and imported to local raster GIS software. Due to the simple raster structure of the model, detailed information on population values and locations is thus obtainable in a highly compact format. Using such data, many conventional analyses may be performed more efficiently, and other new analytical techniques are possible which conventional data structures could not support. These new techniques include the direct estimation of population totals for non-standard areal units, new techniques for the mapping and evaluation of incidence rates, and the identification and analysis of discrete settlements and neighbourhoods. Plate 8 shows the same region as Plate 5, and has been produced by identifying and grouping adjacent populated cells in order to extract each discrete settlement as a single spatial object in the output map. The map in Figure 9.6 shows extremes of population change in the Cardiff region, derived from 1971 and 1981 Census data. These data

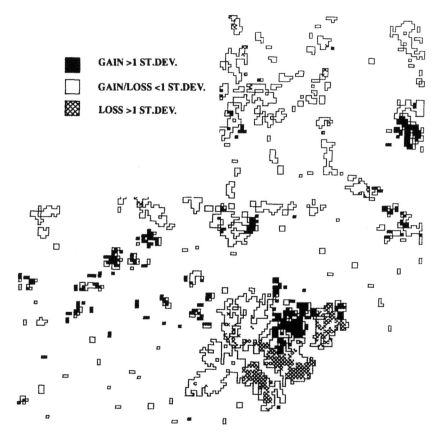

Figure 9.6 Population change in the Cardiff region 1971–1981 (© Crown Copyright)

180

have incompatible zonal geographies, but regions of gain and loss are clearly revealed by surface-based mapping. The construction of surface models from such datasets facilitates direct analysis of both geographic and attribute characteristics. Examples of the cartographic modelling techniques which may be applied to these data were discussed on pp. 122–127. The data model incorporates a high degree of truly geographic information about the population, thus allowing analyses which are meaningful in terms of geographic concepts such as distance between settlements, catchment regions, and variable population density.

Conventional forms of analysis which are readily performed on any surface-based model include the generation of multivariate indices, such as neighbourhood classifications, and the calculation of incidence rates for point-referenced events. This is possible using a neighbourhood function which moves a window across the raster model, centred on each cell in turn. Incidence rates are calculated for each window location and assigned to the central cell. This avoids the need to associate point-referenced events with specific areal units, and efficiently avoids a series of complex coordinate calculations. Population and related counts may be estimated for non-standard areal units by overlaying rasterized zones on the surface matrix and calculating the count value falling in each zone. This is effectively the second stage of areal interpolation in which the surface construction formed the first stage, and the cells of the surface model are the intermediate zones. Additionally, discrete settlements may be uniquely identified as clusters of populated cells surrounded by an unpopulated region, even though this information is not explicitly present in the input data. The implementation of these techniques avoids the need to make unrealistic assumptions about population geography and density, as these characteristics are modelled and present in the database.

The use of raster structures to represent such surfaces has the special property that an estimated density for the area of each cell is also a map of the estimated population in each cell. No sophisticated algorithms (e.g. contour interpolation and storage) are required in the display and modelling of such a map, as the surface-construction algorithm provides estimates of the surface value at all points which are to be displayed (the grid points). Thus in this case a continuous raster map is a choropleth map with very many uniformly sized and shaped areas: the construction of such surfaces is therefore a special type of areal interpolation and it should be seen at one end of a continuum rather than as a completely separate approach from those already discussed. Such models are easily manipulated by existing GIS software and their construction does not require extensive digitizing. Each population-derived variable may have its own geography, irrespective of zonal boundaries, and the data structure is easily able to cope with unpopulated cells. All such models suffer from the important disadvantage of unfamiliarity and imprecision: users are usually most happy to deal with data in conventional formats, and the grid-based surfaces do not meet this requirement. In addition, the modelled aspect of the dataset is made highly visible because the population values in each cell are only estimates, and the modelling process must be explained to the user. The need to describe the method of surface construction is an additional obstacle in illustrating the utility of such approaches to GIS users. In reality, all representations of socioeconomic phenomena are to some degree arbitrarily imposed models, but there is a tendency to forget the restrictions of the most widely used data types.

SUMMARY

Chapter 4 presented a theoretical framework for GIS which focused on the processes acting on spatial data in its passage through the information system. In the light of this framework, the problems commonly associated with the representation of population data are seen to be rooted in the nature of the transformations undergone during data collection and aggregation. The continued use of areal aggregate data for spatial manipulation and analysis is only one of the available design options for the construction of a socioeconomic GIS. There is therefore a need for alternative strategies

Table 9.1 Georeferencing of population: examples of different techniques

Level of aggregation	Example objects	Example products or techniques
Individual (household)	All addresses	PAC directory ADDRESS-POINT Property registers
Geographic aggregation	Data collection zones	Census zones
	Thiessen polygons from centroids	ED centroids
	Dasymetric mapping	Land-use polygons with census data
	Aggregate data at zone centroids	Postcode locations
	Raw data aggregated to grid cells	1971 UK Census
Modelled representation	Based on dasymetric approach	Langford and Unwin, 1994
	Reallocation from zone centroids	Martin and Bracken, 1991
	TIN-based models	Sadler and Barnsley, 1990
	Area interpolation approaches:	
	(i) using control zones	Goodchild *et al.*, 1993a
	(ii) using ancillary data	Flowerdew and Green, 1991

for the representational modelling of socioeconomic phenomena which are at the heart of GIS operations.

A variety of different approaches to modelling socioeconomic phenomena in GIS have been compared here, and a summary of these examples is given in Table 9.1. The examples cited in the table represent different methods for attaching population values to geographic coordinates and do not include the various display techniques such as cartograms and graphical rational mapping. As users have become aware of the power of GIS technology as a tool for handling the growing quantities of spatially referenced information, so dissatisfaction with conventional area-based data models has grown. Alternatives have been sought either through increased data precision or through the new modelling techniques available in GIS environments. We have observed the growth in the use of geodemographic techniques, and Hopwood (1989) notes the rapid expansion of GIS exploitation in the commercial sector as a marketing tool, drawing on many of the methods described in this book. There is clearly a demand for well-designed GIS implementations for socioeconomic information.

We have seen how population information may be represented as point-, area- or surface-type objects, and identified some of the main weaknesses of each approach. The real-world phenomena are actually mobile point-referenced events, but these cannot be adequately captured and handled as spatial data. Areal aggregations of such data lead to the loss of spatial form and difficulties in interpretation associated with the modifiable areal unit problem and the ecological fallacy, due to the unknown and complex nature of the relationship between the areal units and the phenomenon under study. It is attractive to think that the attribute values associated with zone-based representations are 'correct' and therefore more suitable for manipulation within a computer database, but this precision relates only to a single set of arbitrarily imposed boundaries.

In the future, it is anticipated that the sophistication of individual-level databases will continue to grow and that these will form the basis for many geographic analyses. Developments such as national individual-address georeferencing will be of massive importance to many applications such as those concerned with utility management and taxation control. Despite these developments, some important datasets will always be produced at area-aggregate level in order to protect individual confidentiality, and these are most likely to be the data of interest to the academic human geographer. It is suggested in the light of this review, that an appropriate complementary approach to the representation of population data within a GIS is the generation of population-density surfaces in all cases where suitable input data are available. These surfaces can only ever be an estimation, because the collected data have been transformed away from their original spatial-object class, but this is the only representation which allows us freedom to manipulate the population model with validity according to the framework outlined. It must be stressed that all data structures within GIS are merely models (and often very subjectively selective ones) of reality, and the ideal of an exact reproduction of what exists in the real world is illusory. Indeed, we might well argue that no commonly acceptable definition of such 'reality' can be achieved. Early surface-construction techniques were severely limited by the way they treated the input areal data, and this led to a failure to preserve population volume and an inability to identify unpopulated areas. However, GIS technology itself has provided suitable tools for the construction of more sophisticated models which are a serious alternative to conventional approaches.

10

CONCLUSION

In Chapter 1 GIS were introduced as a particular type of automated information system concerned with the management of data relating to specific locations. These systems have been portrayed as the latest stage in a long line of development of methods for representing maps and mappable information. Their current state of development offers enormous power for answering questions relating to geographic location, and they are consequently applicable to a very wide range of operational problems and research interests. The packaging of sophisticated spatial manipulation algorithms in readily accessible software has led to a continuing explosion of interest in GIS, and this has acquired a breadth and momentum far greater than that achieved by earlier techniques for geographic data handling. We have now reviewed the development and techniques of GIS, and have given particular attention to the ways in which such systems may be used by those concerned with the representation of the socioeconomic environment. In this Conclusion it is necessary to reconsider some of the key issues concerning GIS and socioeconomic data which have been raised, and to draw together the many diverse strands which are a feature of any discussion of GIS.

GEOGRAPHIC INFORMATION SYSTEMS . . .

The most important feature of GIS as characterized here, which has the potential to make them more than merely a 'toolbox' of algorithms for geographic data manipulation, is the presence of a dynamic 'model' of the geographic world. The abstraction and representation of certain aspects of the 'real' world by digital means is common to many types of information system, where it provides an environment for queries and experimentation which would be expensive or impractical to perform in reality. It must nevertheless be recognized that the assumption that a single objective 'reality' exists and can be measured is a naive view, and that aspects of the choice of data for incorporation into the system are always either subjective or arbitrary. A variety of different data structures are available in GIS, which allow the construction of representations of geographic objects according to measurements of some of their characteristics, including both location in space and other non-spatial attributes. The values of these measurements are not fixed, as in the symbols of a traditional map, but may be selectively adjusted and extracted to produce alternative representations, or combined to produce new information, which was not necessarily a part of the input data. The different structures which exist for modelling in this way result from different concepts of geographic reality, and the corresponding way to collect and store its characteristics. Modern GIS software provides an integrated environment in which the user can define and employ complex operations on these structures.

It may be argued that most of the techniques found in GIS (such as spatial interpolation methods and the representation of a geographic variable by values in a georeferenced grid) have a long history of development in quantitative geography, and GIS thus represent the drawing-together of many pre-existing methods (e.g. Tobler, 1967; Lam, 1983; Sui, 1994). The actual software systems presently available have strong links with systems for computer-assisted cartography (especially vector-based systems) and remotely sensed image processing (raster-based), which share a concern with the representation of geographic phenomena. Not only are GIS found at the meeting point of different types of information system, but they are proving of interest to agencies concerned with many different application fields and to specialists in many academic disciplines. It is notable that the advent of GIS has prompted many to think of their objects of study in geographic terms when they would not previously have done so. The greatest number of applications have been concerned with some aspect of the physical environment, for example land resources or utility and infrastructure management. Some of the most influential early systems for population mapping such as the North American DIME system used data structures and techniques which were closely allied to those used in physical applications. It has been noted that specific application requirements have often prompted development of new GIS features. Advances in hardware capabilities have also played a significant role, from the limitations imposed by early line-printer mapping to the power available in an integrated workstation environment. All these influences have served to deny the role of any clearly defined theoretical considerations in guiding GIS development or understanding its operation.

Most attempts to develop theories for GIS have been strongly influenced by software structure. These have focused on the basic operations performed or the main components of large systems, and broadly correspond to the modular structure of conventional GIS programs. Sound theoretical work should be able to offer a framework into which actual systems can be placed, and to guide new applications. The existing formulations are largely descriptive and do not offer direction for the application and evaluation of

GIS in new contexts, except as a checklist of software functionality. As new approaches to GIS are developed using object-oriented structures and expert-systems concepts, for example, these theoretical structures will become increasingly unhelpful in understanding GIS operation. In addition, software-oriented concepts of GIS functionality and operation offer little help in terms of defining an appropriate role for GIS: are these systems merely sophisticated tools for the extension of empirical research and organizational management or do they have a more fundamental position in the conduct of geography as an academic discipline?

A more consistent view has been suggested which is developed from a theoretical approach to cartography (Clarke, 1990), just as GIS may be seen as a development from traditional cartographic processing (Figures 4.3 and 4.4). This view is based on the transformations undergone by geographic data within the system, and focuses attention on the nature of the data model, already identified as one of the most important features of GIS. The four data-transformation stages identified in this way have been used as a basis for the discussion of GIS techniques in Chapters 5 to 8. As a result of the strong influence of technical issues on thinking about GIS, the distinction between vector and raster modes of representation has been over-emphasized in the literature, but this distinction may now be seen as less significant. Both are methods (and not the only, or necessarily the best ones) for structuring data, which have traditionally been associated with certain types of application. Although there are computational and storage implications which follow from the decision to use vector or raster approaches, it is argued that the most important issue is to use a data model which closely represents the user's concept of the geographic objects to be represented. There are other data structures such as TIN and object-oriented models which in some cases will be more appropriate than either of the traditional approaches. It is not unreasonable to expect that increased experience with GIS will lead to the development of further methods for representing spatial phenomena (for example, a method based on hierarchical tessellation is suggested by Dutton, 1989). Any informed use of GIS will recognize that each of the transformations (collection, input, manipulation and output), will be subject to the value-systems of the system designer and user, and that many existing applications serve to reinforce the already-powerful in society. This focus on the characteristics of geographic objects and the most appropriate means for their representation particularly prompts a reconsideration of the methods used for the handling of socioeconomic data within GIS. Chapter 9 outlined a variety of different ways in which existing GIS methods may be used in the representation of population-based data. Due to the unique characteristics of population and its attributes, the same structures cannot be expected to handle human and physical phenomena equally well. Another feature of the essentially technology-oriented view of GIS has been a failure to adequately model and interpret error in spatial information, although awareness of the processes of error is now becoming more widespread and techniques are being developed to measure and control the complex processes of error propagation present in all geographic data manipulation. A useful collection of these techniques may be found in Goodchild and Gopal (1989).

GIS are undoubtedly powerful tools, but their implications reach far beyond the mere automation of tasks previously performed by other means. At its best, the ability to model and manipulate geographic information in so many ways poses new questions for spatial analysis and will open the way to new spatial-analytical techniques (as illustrated by Openshaw *et al.*, 1987). A promising sphere of application lies in the need for realistic integration of human and environmental data, fuelled by the growing concern for environmental issues. There is, however, a danger in seeking a 'GIS solution' to all types of problem. There will be many situations in which spatial information provides only part of the answer and many others in which successful application will require more than the direct transfer of different datasets into existing GIS technology. If the current growth in system purchases is to prove worthwhile, there is a need for corresponding growth in attention to theory, data models and quality, and user education, which have major organizational implications for GIS-using agencies. Without these advances, many applications will prove unsatisfactory, and GIS will be seen as having failed to live up to initial expectations, a criticism which has also been levelled at remote sensing, early computer mapping and quantitative methods in geography. It is against this background of organizational and theoretical adjustment, lagging rather behind rapid technical advances, that we must consider attempts to use GIS to model the socioeconomic environment.

. . . AND THEIR SOCIOECONOMIC APPLICATIONS

The most basic requirement for the growth of socioeconomic applications for GIS technology is the supply of large volumes of suitable spatially referenced data, and this is rapidly developing. Most GIS users in this field will be unable to directly collect the data they require, and thus access to secondary databases held by other agencies is essential for the construction of useful information systems. With new censuses in many countries at the start of the 1990s and increasing use of postal systems as georeferences on many other population-related datasets, the collection of information does not appear to be a constraint. An issue of increasing importance will be the availability of these datasets. As illustrated in Chapter 7, many GIS operations themselves add commercial value to basic socioeconomic data, and there is thus a strong incentive for agencies to withhold data for commercial advantage or financial gain. The role of governments is particularly significant here, and different policies are evident in different countries. Despite the technological developments which have made networking solutions possible for most data-transfer requirements, the issue of data access is far from resolved. W. Smith (1990) notes that many of the obstacles to free data interchange may be due to manpower, financial or organizational problems rather than genuine data confidentiality. Rhind (1994) argues strongly that governments should seek to recover from users the costs incurred in the collection and maintenance of national digital databases.

The release of new census data has served to fuel the development of new families of GIS-based software for a wide range of 'geodemographic' tasks including marketing, health-care and other service delivery management, and environmental monitoring. Despite the role of censuses as a catalyst, the dominant trend here is of customization of systems for specific applications and the use of proprietary or in-house socioeconomic data. The importance of these data, rather than the systems themselves, will secure GIS an important place in decision-making processes. The increasing use of GIS as tools enabling managers to plan and evaluate the delivery of goods and services seems inevitable in an era in which any technology which permits tighter control over resource allocation is eagerly adopted. Their role in providing new insights into our understanding of socioeconomic conditions over space, beyond those concerned with the simple reporting of patterns, remains relatively underdeveloped.

Additional contemporary issues relating to socioeconomic data are confidentiality and the choice of basic spatial units. These issues relate not to the techniques used for data processing, but to the characteristics of the underlying 'model'. We have already stressed the need to carefully consider the way in which the model at the heart of all this activity reflects the reality which we are seeking to understand. Although many organizations hold and use personal data, there is an understandable reluctance to allow the free interchange of this kind of information. The influence of the user's values in data collection and measurement decisions will be particularly great in many socioeconomic applications where the data are to be used for social or financial control. For many applications, aggregate or modelled data will offer the only practical basis for the representation of socioeconomic information in GIS (Chapter 9). The most common approach has been to use aggregation-based systems. With these, one of the most significant decisions is that concerning the basic spatial units to be used (Openshaw, 1990). Postal geography is the most widely used form of georeferencing, although it lacks precise spatial definition at the lowest level. All such zoning schemes are hampered to some extent by the modifiable areal unit problem and the ecological fallacy (Openshaw, 1984), which present fundamental difficulties to spatial analysis. New address-level georeferencing will be an important development for many government and commercial applications and is probably more significant than any piece-meal improvements in the geographic referencing of aggregate datasets.

An important question is whether or not this explosion in socioeconomic GIS applications is actually 'good GIS'. The existing theoretical approaches, with their essentially descriptive and technical basis, are unable to address this issue, as they do not seek to explain the processes undergone by spatial information or to define appropriate data models. It is suggested in the light of the review presented here that aggregation-based data structures are severely restrictive as a basis for GIS representation. By contrast, the use of modelled databases offers a flexible and consistent structure for the representation of the socioeconomic environment. There is a long history of attempts to model socioeconomic phenomena, and GIS provides an environment in which old and new ideas may be practically applied. Sur-face-based models, for example, may be generated from existing aggregate data, and more closely reproduce the characteristics of the real-world phenomena. GIS techniques are available for their manipulation and analysis and, additionally, there is the potential to perform a new range of operations not possible with previously existing systems. The incorpora-tion of 'fuzzy' concepts and the development of innovative spatial analyses are more readily incorporated into such approaches.

The general importance of organizational changes, noted above, is perhaps even more significant in the context of population-related data handling. Heywood (1990) notes the influential role of government in determining the atmosphere in which data are exchanged and made avail-able, reinforcing the argument for free circulation of ideas and data in the GIS field. Widespread interchange of spatial data in turn raises difficult questions about ownership and copyright of geographic information and the legal implications of its use. GIS will undoubtedly have a lasting impact on all aspects of socioeconomic data handling, and this is becoming apparent as plans are laid for future censuses and other population-related datasets. National statistical organizations are now looking to GIS solutions to the organization of their data-collection strategies into the next millen-nium. Software and hardware environments will continue to benefit from rapid increases in power and decreases in relative cost. There are still many organizations which are yet to realize the potential of geodemographic analyses for their particular activity. However, the geographic quality of these applications and the uses to which they can realistically be put depends on a fundamental analysis of the way in which socioeconomic phenomena are conceptualized and the selection of appropriate data models. This requires more adequate theories and a willingness on the part of users to break away from conventional cartographic products.

We have seen that it is no longer the technical, but the organizational aspects of GIS which will determine their future. There is also an increasing gulf between the commercial and broader academic concerns surrounding

GIS development. There is now a well-established worldwide industry concerned with the sale and implementation of hardware and software for GIS. For many commercial applications, the arrival of higher-resolution databases and new data structures will make management tasks more efficient, and the benefits will be seen in the daily operation of organizations. There is still a great need, however, for academic enquiry both regarding the real role of GIS and in research into socioeconomic geography. GIS education must not be reduced to training in the use of tools, however powerful, but must develop understanding of their significance and appropriate use. There is clearly a need for continued development of technology, in terms of new data models, understanding of error, and major issues such as temporal GIS. In addition, there is surely a very real contribution which can be made to empirical research in human geography, changing the focus of activity from the systems themselves to the investigation of the world which they were intended to represent.

BIBLIOGRAPHY

Abler, R. F. (1987) 'The National Science Foundation Center for Geographic Information and Analysis' *International Journal of GIS* 1, 4: 303–326

Abler, R. F., Adams, S. and Gould, P. (1971) *Spatial Organization* Englewood Cliffs NJ: Prentice-Hall International

Alper Systems (1990) 'Staying ahead of the game' *Mapping Awareness* 4, 4: 3–8

American Farmland Trust (1985) *A Survey of Geographic Information Systems – for Natural Resources Decision Making at the Local Level* Washington DC: AFT

Anselin, L., Dodson, R. F. and Hudak, S. (1993) 'Linking GIS and spatial data analysis in practice' *Geographical Systems* 1, 1: 3–23

Arbia, G. (1989) 'Statistical effect of spatial data transformations: a proposed general framework' in M. F. Goodchild and S. Gopal (eds) *Accuracy of Spatial Databases* London: Taylor and Francis

Arnaud, A. M. (1993) 'GIS in Portugal: from astrolabe to GPS' *GIS Europe* 2, 2: 11–12

Arnaud, A. M., Craglia, M., Masser, I., Salge, F. and Scholten, H. (1993) 'The research agenda of the European Science Foundation's GISDATA scientific programme' *International Journal of GIS* 7, 1: 463–470

Aronson, P. (1987) 'Attribute handling for geographic information systems' in N. R. Chrisman (ed.) *Proceedings Auto Carto 8* 346–355

Avery, T. E. and Berlin, G. L. (1992) *Fundamentals of Remote Sensing and Airphoto Interpretation* 5th edn, New York: Macmillan

Aybet, J. (1994) 'The object-oriented approach: what does it mean to GIS users?' *GIS Europe* 3, 3: 38–41

Bachi, R. (1968) *Graphical Rational Patterns: A New Approach to Graphical Presentation of Statistics* Jerusalem: Israel Universities Press

Bachi, R. and Hadani, S. E. (1988) 'A package for rational computerized mapping and geostatistical analysis and parameters of distributions of populations' Paper presented at the joint meeting of the Quantitative and Medical study groups of the Institute of British Geographers, Lancaster University

Banai, R. (1993) 'Fuzziness in geographical information systems: contributions from the analytic hierarchy process' *International Journal of GIS* 7, 4: 315–330

Barr, B. (1993a) 'Census Geography II: a review' in A. Dale and C. Marsh (eds) *The 1991 Census User's Guide* London: HMSO

—————— (1993b) 'Mapping and spatial analysis' in A. Dale and C. Marsh (eds) *The 1991 Census User's Guide* London: HMSO

Barrett, E. C. and Curtis, L. F. (1992) *Introduction to Environmental Remote Sensing* London: Chapman and Hall

Batty, M. and Longley, P. (1994) *Fractal Cities* London: Academic Press

Beaumont, J. (ed.) (1989) 'Theme papers on market analysis' *Environment and Planning A* 21, 5
———— (1991) *An Introduction to Market Analysis* Concepts and Techniques in Modern Geography 53, Norwich: Environmental Publications
Bell, J. F. (1978) 'The development of the Ordnance Survey 1 : 25 000 scale derived map' *The Cartographic Journal* 15: 7–13
Berry, B. J. L. and Baker, A. M. (1968) 'Geographic sampling' in B. J. L. Berry and D. F. Marble (eds) *Spatial Analysis: A Reader in Statistical Geography* Englewood Cliffs NJ: Prentice-Hall
Berry, J. K. (1982) 'Cartographic Modelling: procedures for extending the utility of remotely sensed data' in B. F. Richason (ed.) *Remote Sensing: An Input to Geographic Information Systems in the 1980s, Proceedings of the Pecora 7 Symposium, Sioux Falls, S. Dakota, 1981* Falls Church VA: American Society of Photogrammetry
———— (1987) 'Fundamental operations in computer-assisted map analysis' *International Journal of GIS* 1, 2: 119–136
Bertin, J. (1981) *Graphics and Graphic Information Processing* Berlin: Walter de Gruyter
———— (1983) 'A new look at cartography' in D. R. F. Taylor (ed.) *Progress in Contemporary Cartography* vol. 2, Chichester: Wiley
Bickmore, D. P. and Tulloch, T. (1979) 'Medical maps' *Proceedings Auto Carto 4* vol. 1: 324–331
Birkin, M. (1995) 'Customer targeting, geodemographics and lifestyle approaches' in P. Longley and G. Clarke (eds) *GIS for Business and Service Planning* Harlow: Longman
Blakemore, M. (1991) 'Managing an operational GIS: the UK National On-line Manpower Information System (NOMIS)' in D. J. Maguire, M. F. Goodchild and D. W. Rhind (eds) *Geographical Information Systems: Principles and Applications* Harlow: Longman
Blalock, M. (1964) *Causal Inferences in Nonexperimental Research* University of North Carolina: Chapel Hill
Boots, B. N. (1986) *Voroni (Thiessen) Polygons* Concepts and Techniques in Modern Geography 45, Norwich: Geo Books
Boursier, P. and Faiz, S. (1993) 'A comparative study of relational, extensible and object-oriented approaches for modelling and querying geographic databases' *Proceedings EGIS 93* Utrecht: EGIS Foundation
Bowman, A. W. (1985) 'A comparative study of some kernel-based non-parametric density estimators' *Journal of Statistical Computation Simulation* 21: 313–327
Boyle, P. and Dunn, D. (1990) 'Redefining enumeration district centroids and boundaries' Research Report 7, North West Regional Research Laboratory, Lancaster University
Bracken, I. and Martin, D. (1989) 'The generation of spatial population distributions from census centroid data' *Environment and Planning A* 21: 537–543
———— (1995) 'Linkage of the 1981 and 1991 Censuses using surface modelling concepts' *Environment and Planning A* 26, 3: 379–390
Bracken, I. and Webster, C. (1989a) *Information Technology for Geography and Planning* London: Routledge
———— (1989b) 'Towards a typology of geographical information systems' *International Journal of GIS* 3, 2: 137–152
Brand, M. J. D. (1988) 'The geographical information system for Northern Ireland' *Mapping Awareness* 2, 5: 18–21
Broome, F. R. and Meixler, D. B. (1990) 'The TIGER database structure' *Cartography and Geographic Information Systems* 17, 1: 39–47
Brown, P. J. B. (1991) 'Exploring geodemographics' in I. Masser and M. Blakemore

(eds) *Handling Geographical Information: Methodology and Potential Applications* Harlow: Longman

Brusegard, D. and Menger, G. (1989) 'Real data and real problems: dealing with large spatial databases' in M. F. Goodchild and S. Gopal (eds) *Accuracy of Spatial Databases* London: Taylor and Francis

Bryant, N. A. and Zobrist, A. L. (1982) 'Some technical considerations on the evolution of the IBIS system' in B. F. Richason (ed.) *Remote Sensing: An Input to Geographic Information Systems in the 1980s, Proceedings of the Pecora 7 Symposium* Falls Church VA: American Society of Photogrammetry

Bundock, M. S. (1987) 'An integrated DBMS approach to geographical information systems' in N. R. Chrisman (ed.) *Proceedings Auto Carto 8* 292–301

Burrough, P. A. (1986) *Principles of Geographic Information Systems for Land Resources Assessment* Monographs on Soil Resources Survey 12, Oxford: Oxford University Press

—— (1994) 'Accuracy and error in GIS' in D. Green and D. Rix (eds) *The AGI Source Book for Geographic Information Systems 1995* London: Association for Geographic Information

Burrough, P. A. and Boddington, A. (1992) 'The UK Regional Research Laboratory Initiative 1987-1991' *International Journal of GIS* 6, 5: 425–440

Burton, M. (1989) 'Pinpointing the charge register' *Municipal Journal,* 27 January

Carstairs, V. and Lowe, M. (1986) 'Small area analysis: creating an area base for environmental monitoring and epidemiological analysis' *Community Medicine* 8, 1: 15–28

Carstensen, L. W. Jr. (1986) 'Regional land information system development using relational databases and geographic information systems' in M. Blakemore (ed.) *Proceedings Auto Carto London* vol. 1: 507–517

Carver, S. J. and Brunsdon, C. F. (1994) 'Vector to raster conversion error and feature complexity: an empirical study using simulated data' *International Journal of GIS* 8, 3: 261–270

Cassettari, S. (1993) *Introduction to Integrated Geo-Information Management* London: Chapman and Hall

CEOS (Committee on Earth Observation Satellites) (1992) *The Relevance of Satellite Missions to the Study of the Global Environment* London: CEOS, British National Space Centre

Chance, A., Newell, R. and Theriault, D. (1990) 'An object-oriented GIS: issues and solutions' in *Proceedings of the First European Conference on Geographic Information Systems* vol. 1, 179–188, Utrecht: EGIS Foundation

Chen, G. (1986) 'The design of a spatial database management system' in M. Blakemore (ed.) *Proceedings Auto Carto London* vol. 1: 423–432

Chernoff, H. (1973) 'The use of faces to represent points in *k*-dimensional space graphically' *Journal of the American Statistical Association* 68, 342: 361–368

Chorley, R. (1988) 'Some reflections on the handling of geographical information' *International Journal of GIS* 2, 1: 3–9

Chorley, R. J. and Haggett, P. (1965) 'Trend surface models in geographical research' *Transactions of the Institute of British Geographers* 37: 47–67

Chrisman, N. R. (1987) 'Efficient digitizing through the combination of appropriate hardware and software for error detection and editing' *International Journal of GIS* 1, 3: 265–277

—— (1991) 'The error component in spatial data' in D. J. Maguire, M. F. Goodchild and D. W. Rhind (eds) *Geographical Information Systems: Principles and Applications* Harlow: Longman

194

Chrisman, N. R. and Niemann, B. J. (1985) 'Alternative routes to a multipurpose cadastre' *Proceedings Auto Carto 7* 84–94

Clarke, K. C. (1990) *Analytical and Computer Cartography* Englewood Cliffs NJ: Prentice-Hall

Collins, S. H., Moon, G. C. and Lehan, T. H. (1983) 'Advances in geographic information systems' *Proceedings Auto Carto 6* vol. 1: 324–334

Congalton, R. G. (1986) 'Geographic information systems specialist: a new breed' *Proceedings of Geographic Information Systems Workshop, Atlanta GA* Falls Church VA: American Society for Photogrammetry and Remote Sensing

Cooke, D. (1989) 'TIGER and the post-GIS era' *GIS World* July/August: 40–55

Coppock, T. and Anderson, E. (1987) Editorial, *International Journal of GIS* 1, 1: 1–2

Cressman, G. P. (1959) 'An operational objective analysis system' *Monthly Weather Review* 87, 10: 367–374

CRU/OPCS/GRO(S) (1980) *People in Britain: A Census Atlas* London, HMSO

Curran, P. J. (1984) 'Geographic information systems' *Area* 16, 2: 153–158

—— (1985) *Principles of Remote Sensing* Harlow: Longman

—— (1993) 'Earth observation comes of age' *GIS Europe* 2, 4: 27–31

Cushnie, J. (1994) 'A British standard is published' *Mapping Awareness* 8, 5: 40-43

Dale, A. and Marsh, C. (eds) (1993) *The 1991 Census User's Guide* HMSO, London

Dangermond, J. (1983) 'A classification of software components commonly used in geographic information systems' in D. J. Peuquet and J. O'Callaghan (eds) *Design and Implementation of Computer-based Geographic Information Systems* Amherst, NY: IGU Commission on Geographical Data Sensing and Processing

Dangermond, J. and Freedman, C. (1987) 'Findings regarding a conceptual model of a municipal database and implications for software design' in W. J. Ripple (ed.) *GIS for Resource Management: A Compendium* Falls Church VA: American Society for Photogrammetry and Remote Sensing

Dansby, H. B. (1994) 'Access to digital data in US federal agencies' in D. Green and D. Rix (eds) *The AGI Source Book for Geographic Information Systems 1995* London: Association for Geographic Information

Date, C. J. (1990) *An Introduction to Database Systems* vol. 1, 5th edn, Reading MA: Addison-Wesley

Dent, B. D. (1985) *Principles of Thematic Map Design* Reading MA: Addison-Wesley

Department of the Environment (1987) *Handling Geographic Information. The Report of the Committee of Enquiry Chaired by Lord Chorley* London: HMSO

—— (1988) *Handling Geographic Information. The Government's Response to the Report of the Committee of Enquiry Chaired by Lord Chorley* London: HMSO

Devereux, B. J. (1986) 'The integration of cartographic data stored in raster and vector formats' in M. Blakemore (ed.) *Proceedings Auto Carto London* vol. 1: 257–266

Dickinson, H. J. and Calkins, H. W. (1988) 'The economic evaluation of implementing a GIS' *International Journal of GIS* 2, 4: 307–327

Diggle, P. H. (1983) *Statistical Analysis of Spatial Point Patterns* London: Academic Press

Dobson, M. W. (1979) 'Visual information processing during cartographic communication' *The Cartographic Journal* 16: 14–20

Dorling, D. (1993) 'Map design for census mapping' *The Cartographic Journal* 30, 2: 167–183

—— (1994) 'Cartograms for visualizing human geography' in H. M. Hearnshaw and D. J. Unwin (eds) *Visualization in Geographical Information Systems* Chichester: Wiley

—————— (1995) 'Visualizing changing social structure from a census' *Environment and Planning A* 26, 3: 353-378

Doytsher, Y. and Shmutter, B. (1986) 'Intersecting layers of information – a computerized solution' in M. Blakemore (ed.) *Proceedings Auto Carto London* vol. 1: 136–145

Drury, P. (1987) 'A geographic health information system' *BURISA* 77: 4–5

Drury, S. A. (1990) *A Guide to Remote Sensing: Interpreting Images of the Earth* Oxford: Oxford University Press

Dueker, K. J. (1985) 'Geographic information systems: toward a geo-relational structure' *Proceedings Auto Carto 7* 172–177

—————— (1987) 'Geographic information systems and computer-aided mapping' *APA Journal* Summer 1987: 383–390

Duggin, M. J., Rowntree, R. A. and Odell, A. W. (1988) 'The application of spatial filtering methods to urban feature analysis using digital image data' *International Journal of Remote Sensing* 9, 3: 543–553

Dutton, G. (1989) 'Modelling locational uncertainty via hierarchical tessellation' in M. F. Goodchild and S. Gopal (eds) *Accuracy of Spatial Data* London: Taylor and Francis

Eastman, J. R. (1992) *IDRISI User's Guide Version 4.0 rev 1* Worcester MA: Clark University Graduate School of Geography

Egbert, S. L. and Slocum, T. A. (1992) 'EXPLOREMAP: an exploration system for choropleth maps' *Annals of the Association of American Geographers* 82, 2: 275

Epstein, E. F. (1991) 'Legal aspects of GIS' in D. J. Maguire, M. F. Goodchild and D. W. Rhind (eds) *Geographical Information Systems: Principles and Applications* Harlow: Longman

ESRI (1993a) *Understanding GIS: The Arc/Info Method* Redlands CA: Environmental Systems Research Institute Inc.

—————— (1993b) *Address Geocoding: Managing Address Information. Arc/Info Users Guide 6.0* Redlands CA: ESRI

Estes, J. E. (1982) 'Remote sensing and geographic information systems coming of age in the eighties' in B. F. Richason (ed) *Remote Sensing: An Input to Geographic Information Systems in the 1980s*, Proceedings of the Pecora 7 Symposium, Falls Church VA: American Society of Photogrammetry

Evans, I. S. (1977) 'The selection of class intervals' *Transactions of the Institute of British Geographers* NS 2: 98–124

Fisher, P. F. (1991) 'Spatial data sources and data problems' in D. J. Maguire, M. F. Goodchild and D. W. Rhind (eds) *Geographical Information Systems: Principles and Applications* Harlow: Longman

Flowerdew, R. and Goldstein, W. (1989) 'Geodemographics in practice, developments in North America' *Environment and Planning A* 21: 605–616

Flowerdew, R. and Green, M. (1991) 'Data integration: statistical methods for transferring data between zonal systems' in I. Masser and M. Blakemore (eds) *Handling Geographical Information: Methodology and Potential Applications* Harlow: Longman

Flowerdew, R. and Openshaw, S. (1987) *A Review of the Problems of Transferring Data from One Set of Areal Units to Another Incompatible Set* Northern RRL Research Report No. 4, Newcastle-upon-Tyne: NRRL

Fotheringham, A. S. and Rogerson, P. A. (1993) 'GIS and spatial analytical problems' *International Journal of GIS* 7, 1: 3–19

Fotheringham, A. S. and Wong, D. W. S. (1991) 'The modifiable areal unit problem in multivariate statistical analysis' *Environment and Planning A* 23, 7: 1025–1034

Fox, T. (1990) 'The GIS plotting decision' *Mapping Awareness* 4, 4: 36–37

Fraser, S. E. (1984) 'Developments at the Ordnance Survey since 1981' *The Cartographic Journal* 21: 59–61

Gardiner, V. and Unwin, D. (1985) 'Limitations of microcomputers in thematic mapping' *Computers and Geosciences* 11, 3: 291–295

Gatrell, A. C. (1988) *Handling Geographic Information for Health Studies* Northern Regional Research Laboratory, Research Report No. 15 Newcastle-upon-Tyne: NRRL

—————— (1989) 'On the spatial representation and accuracy of address-based data in the United Kingdom' *International Journal of GIS* 3, 4: 335–348

—————— (1991) 'Concepts of space and geographic data' in D. J. Maguire, M. F. Goodchild and D. W. Rhind (eds) *Geographical Information Systems: Principles and Applications* Harlow: Longman

Gilbert, C. (1994) 'Portable GPS for mapping: features versus benefits' *Mapping Awareness* 8, 2: 26–29

Goh, P-C. (1989) 'A graphic query language for cartographic and land information systems' *International Journal of GIS* 3, 3: 245–255

Goodchild, M. F. (1984) 'Geocoding and geosampling' in G. L. Gaile and C. J. Willmott (eds) *Spatial Statistics and Models* Dordrecht: D. Reidel

—————— (1987) 'A spatial analytical perspective on geographical information systems' *International Journal of GIS* 1, 4: 327–334

—————— (1991a) 'The technological setting of GIS' in D. J. Maguire, M.F. Goodchild and D. W. Rhind (eds) *Geographical Information Systems: Principles and Applications* Harlow: Longman

—————— (1991b) 'Geographic information systems' *Progress in Human Geography* 15, 2: 194–200

—————— (1992) 'Geographical information science' *International Journal of GIS* 6, 1: 31–45

—————— (1994) 'Information highways' in D. Green and D. Rix (eds) *The AGI Source Book for Geographic Information Systems 1995* London: Association for Geographic Information

Goodchild, M. F. and Dubuc, O. (1987) 'A model of error for choropleth maps, with applications to geographic information systems' in N. R. Chrisman (ed.) *Proceedings Auto Carto 8* 165–174

Goodchild, M. F. and Gopal, S. (eds) (1989) *Accuracy of Spatial Data* London: Taylor and Francis

Goodchild, M. F., Haining, R., Wise, S. and 12 others (1992) 'Integrating GIS and spatial data analysis: problems and possibilities' *International Journal of GIS* 6, 5: 407–423

Goodchild, M. F., Anselin, L. and Deichmann, U. (1993a) 'A framework for the areal interpolation of socioeconomic data' *Environment and Planning A* 25: 383–397

Goodchild, M. F.,Parks, B. O. and Steyaert, L. T. (1993b) (eds) *Environmental Modelling with GIS* Oxford: Oxford University Press

Gould, M. I. (1992) 'The use of GIS and CAC by health authorities: results from a postal questionnaire' *Area* 24, 4: 391–401

Green, D. R. and Rix, D. (eds) (1994) *The AGI Source Book for Geographic Information Systems 1995* London: Association for Geographic Information

Green, N. P. (1987) *Database Design and Implementation* SERRL Working Report No.2, London: SERRL

Green, N. P., Finch, S. and Wiggins, J. (1985) 'The "state of the art" in geographical information systems' *Area* 17, 4: 295–301

Green, P. J. and Sibson, R. (1978) 'Computing Dirichlet tessellations in the plane' *Computer Journal* 21, 2: 168–173

Gunson, G. (1986) 'Welsh Water beats the information shortage' *Surveyor* 21 August: 18

Guptill, S. C. (1989) 'Inclusion of accuracy data in a feature-based, object-oriented data model' in M. F. Goodchild and S. Gopal (eds) *Accuracy of Spatial Databases* London: Taylor and Francis

———— (1991) 'Spatial data exchange and standardization' in D. J. Maguire, M. F. Goodchild and D. W. Rhind (eds) *Geographical Information Systems: Principles and Applications* Harlow: Longman

Haggett, P. (1983) *Geography: A Modern Synthesis* Rev. 3rd edn, New York: Harper and Row

Haines-Young, R. and Green, D. R. (1993) *Landscape Ecology and GIS* London: Taylor and Francis

Harley, J. B. (1989) 'Deconstructing the map' *Cartographica* 26, 2: 1–20

Harris, R. (1987) *Satellite Remote Sensing: An Introduction* London: Routledge and Kegan Paul

Harts, J. J., Ottens, H. F. L. and Scholten, H. J. (1990) 'EGIS '90 and the development and application of geographical information systems in Europe' *Proceedings of the First European Conference on Geographical Information Systems* Utrecht: EGIS Foundation

Harvey, D. (1969) *Explanation in Geography* London: Edward Arnold

Hawkins, D. M. and Cressie, N. (1984) 'Robust kriging – a proposal' *Mathematical Geology* 16: 3–18

Healey, R. G. (1988) 'Interfacing software systems for geographical information systems' *ESRC Newsletter* 63: 23–26

———— (1991) 'Database management systems' in D. J. Maguire, M. F. Good-child and D. W. Rhind (eds) *Geographical Information Systems: Principles and Applications* Harlow: Longman

Hearnshaw, H. M. and Unwin, D. J. (eds) (1994) *Visualization in Geographical Information Systems* Chichester: Wiley

Herring, J. R. (1987) 'TIGRIS: Topologically Integrated Geographic Information System' in N. R. Chrisman (ed.) *Proceedings Auto Carto 8* 282–291

Heywood, I. (1990) 'Commentary: Geographic information systems in the social sciences' *Environment and Planning A* 22: 849–854

Hirschfield, A., Brown, P. and Bundred, P. (1993) 'Doctors, patients and GIS' *Mapping Awareness* 7, 9: 9–12

Hodgson, M. E. (1985) 'Constructing shaded maps with the DIME topological structure: an alternative to the polygon approach' *Proceedings Auto Carto 7* 275–282

Holroyd, F. (1988) 'Image data structures and image processing algorithms' Paper presented to the NERC conference on GIS, Keyworth: NERC

Hopkins, A. and Maxwell, R. (1990) 'Contracts and quality of care' *British Medical Journal* 300: 919–922

Hopwood, D. (1989) 'GIS as a marketing tool' *Mapping Awareness* 3, 3: 28–30

Hoyland, G. and Goldsworthy, D. D. (1986) 'The development of an automated mapping system for the electricity distribution system in the South Western Electricity Board' in M. Blakemore (ed.) *Proceedings Auto Carto London* vol. 2: 171–180

Hume, J. (1987) 'Postcodes as reference codes' *Applications of Postcodes in Locational Referencing* London: LAMSAC

Ingram, K. J. and Phillips, W. W. (1987) 'Geographic information processing using a SQL-based query language' in N. R. Chrisman (ed.) *Proceedings Auto Carto 8* 326–335

Innes, J. E. and Simpson, D. M. (1993) 'Implementing GIS for planning: lessons from the history of technological innovation' *Journal of the American Planning Association* 59, 2: 230-236

Jackson, M. J. and Woodsford, P. A. (1991) 'GIS data capture hardware and software' in D. J. Maguire, M. F. Goodchild and D. W. Rhind (eds) *Geographical Information Systems: Principles and Applications* Harlow: Longman

Jarman, B. (1983) 'Identification of underprivileged areas' *British Medical Journal* 286, 6379: 1705–1709

Jianya, G. (1990) 'Object-oriented models for thematic data management in GIS' *Proceedings of the First European Conference on Geographical Information Systems* Utrecht: EGIS Foundation

Jungert, E. (1990) 'A database structure for an object-oriented raster-based geographic information system' *Proceedings of the First European Conference on Geographical Information Systems* Utrecht: EGIS Foundation

Kehoe, B. P. (1994) *Zen and the Art of the Internet* 3rd edn, London: Prentice-Hall

Kennedy, S. (1989) 'The small number problem and the accuracy of spatial databases' in M. F. Goodchild and S. Gopal (eds) *Accuracy of Spatial Databases* London: Taylor and Francis

Kennedy, S. and Tobler, W. R. (1983) 'Geographic interpolation' *Geographical Analysis* 15, 2: 151–156

Kent, M., Jones, A. and Weaver, R. (1993) 'Geographical information systems and remote sensing in land use planning: an introduction' *Applied Geography* 13, 1: 5–8

Kevany, M. J. (1983) 'Interactive graphics systems in local government mapping: the achievements and challenges' *Proceedings Auto Carto 6* vol. 1: 283–290

Kinnear, C. (1987) 'The TIGER structure' in N. R. Chrisman (ed.) *Proceedings Auto Carto 8* 249–257

Kleiner, A. and Brassel, K. (1986) 'Hierarchical grid structures for static geographic data bases' in M. Blakemore (ed.) *Proceedings Auto Carto London* vol. 1: 485–496

Kollias, V. J. and Voliotis, A. (1991) 'Fuzzy reasoning in the development of geographical information systems. FRSIS: a prototype soil information system with fuzzy retrieval capabilities' *International Journal of GIS* 5, 2: 209–224

Korner, E. (1980) 'The first report to the Secretary of State on the collection and use of information about hospital clinical activity in the National Health Service' Steering Group on Health Services Information in GB, London: HMSO/DHSS

Lam, A. H. S. (1985) 'Microcomputer mapping systems for local governments' *Proceedings Auto Carto 7*: 327–336

Lam, N. S. (1983) 'Spatial interpolation methods: a review' *American Cartographer* 10, 2: 129–149

Land, N. (1989) 'Transfer standards' *Mapping Awareness* 3, 1: 42–43

Langford, M. and Unwin, D. J. (1994) 'Generating and mapping population density surfaces within a geographical information system' *The Cartographic Journal* 31: 21–26

Langram, G. (1992) *Time in Geographic Information Systems* London: Taylor and Francis

Laurini, R. and Thompson, D. (1992) *Fundamentals of Spatial Information Systems* APIC Series 37, London: Academic Press

Lemmens, J. P. M. (1990) 'Strategies for the integration of raster and vector data' *Proceedings of the First European Conference on Geographical Information Systems* Utrecht: EGIS Foundation

Ley, R. G. (1986) 'Accuracy assessment of digital terrain models' in M. Blakemore (ed.) *Proceedings Auto Carto London* vol. 1: 455–464

Lo, C. P. (1986) *Applied Remote Sensing* Harlow: Longman

────── (1989) 'A raster approach to population estimation using high-altitude aerial and space photographs' *Remote Sensing of Environment* 27: 59–71

Logan, T. L. and Bryant, N. A. (1987) 'Spatial data software integration: merging CAD/CAM/Mapping with GIS and image processing' *Photogrammetric Engineering and Remote Sensing* 53, 10: 1391–1395

McDowell, T., Meixler, D., Rosenson, P. and Davis, B. (1987) 'Maintenance of geographic structure files at the Bureau of the Census' in N. R. Chrisman (ed.) *Proceedings Auto Carto 8* 264–269

MacEachran, A. M. and Davidson, J. V. (1987) 'Sampling and isometric mapping of continuous geographic surfaces' *American Cartographer* 14, 4: 299–320

Maffini, G. (1987) 'Raster versus vector data encoding and handling: a commentary' *Photogrammetric Engineering and Remote Sensing* 53, 10: 1397–1398

────── (1990) 'The role of public domain databases in the growth and development of GIS' *Mapping Awareness* 4, 1: 49–54

Maffini, G., Arno, M. and Bitterlich, W. (1989) 'Observations and comments on the generation and treatment of error in digital GIS data' in M. F. Goodchild and S. Gopal (eds) *Accuracy of Spatial Databases* London: Taylor and Francis

Maguire, D. J. (1989) *Computers in Geography* Harlow: Longman

────── (1991) 'An overview and definition of GIS' in D. J. Maguire, M. F. Goodchild and D. W. Rhind (eds) *Geographical Information Systems: Principles and Applications* Harlow: Longman

Maguire, D. J. and Dangermond, J. (1994) 'Future GIS technology' in D. Green and D. Rix (eds) *The AGI Source Book for Geographic Information Systems 1995* London: Association for Geographic Information

Maguire, D. J., Goodchild, M. F. and Rhind, D. W. (eds) (1991) *Geographical Information Systems: Principles and Applications* Harlow: Longman

Mahoney, R. P. (1991) 'GIS and utilities' in D. J. Maguire, M. F. Goodchild and D. W. Rhind (eds) *Geographical Information Systems: Principles and Applications* Harlow: Longman

Marble, D. F. and Peuquet, D. J. (1983) 'Geographic information systems and remote sensing' in R. N. Colwell (ed.) *Manual of Remote Sensing* 1: 923-958

Mark, D. M. (1986) 'The use of quadtrees in geographic information systems and spatial data handling' in M. Blakemore (ed.) *Proceedings Auto Carto London* vol. 1: 517–526

Marshall, C. (1994) 'Survey of publications on GIS, remote sensing and cartography. Part 1: books on GIS' *Mapping Awareness* 8, 7: 40–44

Martin, D. (1989) 'Mapping population data from zone centroid locations' *Transactions of the Institute of British Geographers* NS 14, 1: 90–97

────── (1992) 'Postcodes and the 1991 Census of Population: issues, problems and prospects' *Transactions of the Institute of British Geographers* NS 17, 3: 350–357

────── (1993) *The 1991 UK Census of Population* Concepts and Techniques in Modern Geography 56, Norwich: Environmental Publications

────── (1995) 'Censuses and the modelling of population in GIS' in P. Longley and G. Clarke (eds) *GIS for Business and Service Planning* Harlow: Longman

Martin, D. and Bracken, I. (1991) 'Techniques for modelling population-related raster databases' *Environment and Planning A* 23, 1065–1079

────── (1994) *1981 and 1991 Population Surface Models: A Guide* CSS 605, University of Manchester Computing Centre, Manchester

Martin, D. and Gascoigne, R. (1994) 'Change and change again: geographical implications for intercensal analysis' *Area* 26, 2: 133–141

Martin, D., Longley, P. and Higgs, G. (1994) 'The use of GIS in the analysis of diverse urban databases' *Computers, Environment and Urban Systems* 18, 1: 55–66

Marx, R. W. (1990) 'The TIGER system: yesterday, today and tomorrow' *Cartography and Geographic Information Systems* 17, 1: 89–97

Mason, D. C., Corr, D. G., Cross, A., Hoggs, D. C., Lawrence, D. H., Petrou, M. and Tailor, A. M. (1988) 'The use of digital map data in the segmentation and classification of remotely-sensed images' *International Journal of GIS* 2, 3: 195–215

Medyckyj-Scott, D. (1994) 'Visualization and human-computer interaction in GIS' in H. M. Hearnshaw and D. J. Unwin (eds) *Visualization in Geographical Information Systems* Chichester: Wiley

Merchant, J. W. (1982) 'Employing Landsat MSS data in land use mapping: observations and considerations' in B. F. Richason (ed.) *Remote Sensing: An Input to Geographic Information Systems in the 1980s. Proceedings of the Pecora 7 Symposium* Falls Church VA: American Society of Photogrammetry

Michalak, W. Z. (1993) 'GIS in land use change analysis: integration of remotely sensed data into GIS' *Applied Geography* 13, 1: 28–44

Moellering, H. (1980) 'Strategies of real-time cartography' *The Cartographic Journal* 17: 12–15

Mohan, J. and Maguire, D. J. (1985) 'Harnessing a breakthrough to meet the needs of health care' *Health and Social Service Journal* 9 May: 580–581

Monmonier, M. S. (1982) *Computer-Assisted Cartography. Principles and Prospects* Englewood Cliffs NJ: Prentice-Hall

—— (1983) 'Cartography, mapping and geographic information' *Progress in Human Geography* 7: 420–427

—— (1991) *How to Lie with Maps* Chicago: University of Chicago Press

Morphet, C. (1993) The mapping of small-area census data – a consideration of the effects of enumeration district boundaries *Environment and Planning A* 25, 9: 1267–1277

Morrison, J. L. (1974) 'Observed statistical trends in various interpolation algorithms useful for first stage interpolation' *The Canadian Cartographer*, 11: 142–149

—— (1980) 'Computer technology and cartographic change' in D. R. F. Taylor (ed.) *Progress in Contemporary Cartography* vol. 1, Chichester: Wiley

Mounsey, H. M. (1991) 'Multisource, multinational environmental GIS: lessons learnt from CORINE' in D. J. Maguire, M. F. Goodchild and D. W. Rhind (eds) *Geographical Information Systems: Principles and Applications* Harlow: Longman

Moxey, A. and Allanson, P. (1994) 'Areal interpolation of spatially extensive variables: a comparison of alternative techniques' *International Journal of GIS* 8, 5: 479–487

Muehrcke, P. (1969) 'Visual Pattern Analysis: A Look at Maps' Unpublished PhD Thesis, University of Michigan

NCGIA (1989) 'The research plan of the National Center for Geographic Information and Analysis' *International Journal of GIS* 3, 2: 117–136

—— (1992) 'National Center for Geographical Information and Analysis: Publications list' *International Journal of GIS* 6, 1: 47–52

Newcomer, J. A. and Szajgin, J. (1984) 'Accumulation of thematic map errors in digital overlay analysis' *American Cartographer* 11, 1: 58–62

Niemann, O. (1993) 'Automated forest cover mapping using Thematic Mapper images and ancillary data *Applied Geography* 13, 1: 86–95

Norcliffe, G. B. (1969) 'On the uses and limitations of trend surface models' *Canadian Geographer* 13: 338–348

Nordbeck, S. and Rystedt, B. (1970) 'Isarithmic maps and the continuity of reference interval functions' *Geografiska Annaler* 52B: 92–123

—————— (1972) *Computer Cartography* Lund: Carl Bloms Boktryckeri

Oakes, J. and Thrift, N. (1975) 'Spatial interpolation of missing data: an empirical comparison of some different methods' *Computer Applications*, 2, 3/4: 335–356

OPCS (1986) *Population and Vital Statistics: Local and Health Authority Area Summary 1984* Fareham: OPCS

—————— (1992a) *1991 Census Definitions Great Britain* CEN 91 DEF, London: OPCS

—————— (1992b) *ED/Postcode Directory: Prospectus* 1991 Census User Guide 26, Fareham: OPCS

—————— (1993) *Report on Review of Statistical Information on Population and Housing (1996-2016)* Occasional Paper 40, London: OPCS

Openshaw, S. (1984) *The Modifiable Areal Unit Problem* Concepts and Techniques in Modern Geography 38, Norwich: Geo Books

—————— (1988) 'An evaluation of the BBC Domesday interactive videodisc as a GIS for planners' *Environment and Planning B* 15: 325–328

—————— (1989) 'Learning to live with errors in spatial databases' in M. F. Goodchild and S. Gopal (eds) *Accuracy of Spatial Databases*, London: Taylor and Francis

—————— (1990) 'Spatial referencing for the user in the 1990s' *Mapping Awareness* 4, 2: 24–29

—————— (1991a) 'A view on the GIS crisis in geography, or, using GIS to put Humpty-Dumpty back together again' *Environment and Planning A* 23: 621–628

—————— (1991b) 'Developing appropriate spatial analysis methods for GIS' in D. J. Maguire, M. F. Goodchild and D. W. Rhind (eds) *Geographical Information Systems: Principles and Applications* Harlow: Longman

—————— (1995) 'Marketing spatial analysis: some prospects and technologies of relevance to marketing' in P. Longley and G. Clarke (eds) *GIS for Business and Service Planning* Harlow: Longman

Openshaw, S. and Taylor, P. J. (1981) 'The modifiable areal unit problem' in N. Wrigley and R. J. Bennett (eds) *Quantitative Geography: A British View* London: Routledge and Kegan Paul

Openshaw, S., Cullingford, D. and Gillard, A. (1980) 'A critique of the national classifications of OPCS/PRAG' *Town Planning Review* 51: 421–439

Openshaw, S., Charlton, M., Wymer, C. and Craft, A. (1987) 'A mark 1 geographical analysis machine for the automated analysis of point data sets' *International Journal of GIS* 1: 335–358

Ordnance Survey (1993) *ADDRESS-POINT User Guide* Southampton: Ordnance Survey

Owen, D. W., Green, A. E. and Coombes, M. G. (1986) 'Using the social and economic data on the BBC Domesday disc' *Transactions of the Institute of British Geographers* NS 11: 305–314

Parrott, R. and Stutz, F. P. (1991) 'Urban GIS applications' in D. J. Maguire, M. F. Goodchild and D. W. Rhind (eds) *Geographical Information Systems: Principles and Applications* Harlow: Longman

Pearman, H. (1993) 'Designing a land and property gazetteer' *Mapping Awareness* 7, 5: 19–21

Petrie, G. (1989) 'Networking for digital mapping' *Mapping Awareness* 3, 3: 9–16

Peucker, T. K. and Chrisman, N. (1975) 'Cartographic data structures' *The American Cartographer* 2: 55–69

Peucker, T. K., Fowler, R. J., Little, J. J. and Mark, D. M. (1978) 'The Triangulated Irregular Network' *Proceedings of the DTM Symposium* Falls Church VA: American Congress on Surveying and Mapping

Peuquet, D. J. (1981a) 'An examination of techniques for reformatting digital cartographic data/ Part 1: the raster-to-vector process' *Cartographica* 18, 1: 34–48

———— (1981b) 'An examination of techniques for reformatting digital cartographic data/ Part 2: the vector-to-raster process' *Cartographica* 18, 3: 21–33

Piwowar, J. M., Ellsworth, F. L. and Dudycha, D. J. (1990) 'Integration of spatial data in vector and raster formats in a geographical information system environment' *International Journal of GIS* 4, 4: 429–444

Plane, D. A. and Rogerson, P. A. (1994) *The Geographical Analysis of Population with Applications to Planning and Business* New York: Wiley

Post Office (1985) *The Postcode Address File Digest* London: The Post Office

Pugh, D. (1992) 'The National Land and Property Gazetteer' *Mapping Awareness* 6, 5: 32–45

Raper, J. F., Rhind, D. W. and Shepherd, J. W. (1992) *Postcodes: The New Geography* Harlow: Longman

Rase, W-D. (1987) 'The evolution of a graduated symbol software package in a changing graphics environment' *International Journal of GIS* 1, 1: 51–66

Rhind, D. W. (1977) 'Computer-aided cartography' *Transactions of the Institute of British Geographers* NS 2, 1: 71–97

———— (1983) 'Mapping census data' in D. W. Rhind (ed.) *A Census Users Handbook* London: Methuen

———— (1986) 'Remote sensing, digital mapping, and geographic information systems: the creation of national policy in the UK' *Environment and Planning C* 4, 1: 91–102

———— (1987) 'Recent developments in geographical information systems in the UK' *International Journal of GIS* 1, 3: 229–242

———— (1988) 'A GIS research agenda' *International Journal of GIS* 2, 1: 23–28

———— (1991) 'Counting the people: the role of GIS' in D. J. Maguire, M. F. Goodchild and D. W. Rhind (eds) *Geographical Information Systems: Principles and Applications* Harlow: Longman

———— (1992) 'Data access, charging and copyright and their implications for geographical information systems' *International Journal of GIS* 6, 1: 13–30

———— (1993) 'Policy on the supply and availability of Ordnance Survey information over the next five years' *Mapping Awareness* 7, 1: 37–41

———— (1994) 'Spatial data from government' in D. Green and D. Rix (eds) *The AGI Source Book for Geographical Information Systems 1995* London: Association for Geographic Information

Rhind, D. W. and Green, N. P. A. (1988) 'Design of a geographical information system for a heterogeneous scientific community' *International Journal of GIS* 2, 3: 171–189

Rhind, D. W. and Mounsey, H. (1986) 'The land and people of Britain: a Domesday record, 1986' *Transactions of the Institute of British Geographers* NS 11: 315–325

Rhind, D. W. and Openshaw, S. (1987) 'The BBC Domesday system: a nation-wide GIS for $4448' in N. R. Chrisman (ed.) *Proceedings Auto Carto 8* 595–603

Rhind, D. W., Adams, T., Fraser, S. E. G. and Elston, M. (1983) 'Towards a national digital topographic database: experiments in mass digitising, parallel processing and the detection of change' *Proceedings Auto Carto 6* vol. 1: 428–437

Rix, D. (1994) 'Recent trends in GIS technology' in D. Green and D. Rix (eds) *The AGI Source Book for Geographic Information Systems 1995* London: Association for Geographic Information

Robbins, R. G. and Thake, J. E. (1988) 'Coming to terms with the horrors of automation' *The Cartographic Journal* 25: 139–142

Robinson, A. (1987) *Elements of Cartography* 5th edn, New York: Wiley

Robinson, A. H. and Petchenik, B. B. (1975) 'The map as a communication system' *The Cartographic Journal* 12, 1: 7–14

Rose, M. T. (1992) *The Internet Message: Closing the Book with Electronic Mail* London: Prentice-Hall

Sacker, D. (1987) 'Value added postcode services. The SIA experience' in *Applications of Postcodes in Locational Referencing* London: LAMSAC

Sadler, G. J. and Barnsley, M. J. (1990) 'Use of population density data to improve classification accuracies in remotely-sensed images of urban areas' *Proceedings of the First European Conference on Geographical Information Systems* Utrecht: EGIS Foundation

Sadler, G. J., Barnsley, M. J. and Barr, S. L. (1991) 'Information extraction from remotely sensed images for urban land analysis' in *Proceedings of the Second European Conference on Geographical Information Systems* vol. 2: 955–964, Utrecht: EGIS Foundation

Sandhu, J. S. and Amundson, S. (1987) '*The Map Analysis Package for the PC*' Buffalo: State University of New York at Buffalo

Schmid, C. F. and MacCannell, E. H. (1955) 'Basic problems, techniques, and theory of isopleth mapping' *Journal of the American Statistical Association* 50: 220–239

Shepard, D. S. (1984) 'Computer mapping: the Symap interpolation algorithm' in G. L. Gaile and C. J. Willmott (eds) *Spatial Statistics and Models* Dordrecht: D. Reidel

Shiryaev, E. E. (1987) *Computers and the Representation of Geographical Data* (Translated from Russian) Chichester: Wiley

Sibson, R. (1981) 'A brief description of natural neighbour interpolation' in V. Barnett (ed.) *Interpreting Multivariate Data* Chichester: Wiley

Siderelis, K. C. (1991) 'Land resource information systems' in D. J. Maguire, M. F. Goodchild and D. W. Rhind (eds) *Geographical Information Systems: Principles and Applications* Harlow: Longman

Smith, D. A. and Tomlinson, R. F. (1992) 'Assessing costs and benefits of geographical information systems: methodological and implementation issues' *International Journal of GIS* 6, 3: 247–256

Smith, P. (1988) 'Developments in the use of digital mapping in HM Land Registry' *Mapping Awareness* 2, 4: 37–41

Smith, W. (1990) 'Mapping Awareness '90. Conference chairman's address' *Mapping Awareness* 4, 2: 11–13

Statistics Canada (1992) *1991 Geography Catalogue* Ottawa: Statistics Canada

Sui, D. Z. (1994) 'GIS and urban studies: positivism, post-positivism, and beyond' *Urban Geography* 15, 3: 258–278

Taylor, P. J. and Overton, M. (1991) 'Further thoughts on geography and GIS' *Environment and Planning A* 23: 1087–1094

Teng, A. T. (1983) 'Cartographic and attribute database creation for planning analysis through GBF/DIME and census data processing' in M. Blakemore (ed.) *Proceedings Auto Carto 6* vol. 2: 348–354

—————— (1986) 'Polygon overlay processing: a comparison of pure geometric manipulation and topological overlay processing' *Proceedings, Second International Symposium on Spatial Data Handling, Seattle, Washington* 102–119

Thrall, G. I. and Elshaw Thrall, S. (1993) 'Commercial data for the business GIS' *Geo Info Systems* July/August 63–68

—————— (1994) 'Business GIS data, part 3: ZIP Plus 4 geocoding' *Geo Info Systems* January 57–60

Tobler, W. R. (1967) 'Of maps and matrices' *Journal of Regional Science* 17, 2 (supplement): 275–280

—— (1973) 'Choropleth maps without class intervals?' *Geographical Analysis* 5: 262–265

—— (1979) 'Smooth pycnophylactic interpolation for geographical regions' *Journal of the American Statistical Association* 74, 367: 519–530

Tobler, W. R. and Kennedy, S. (1985) 'Smooth multidimensional interpolation' *Geographical Analysis* 17, 3: 251–257

Tomlin, C. D. (1990) *Geographic Information Systems and Cartographic Modelling* Englewood Cliffs NJ: Prentice-Hall

—— (1991) 'Cartographic modelling' in D. J. Maguire, M. F. Goodchild and D. W. Rhind (eds) *Geographical Information Systems: Principles and Applications* Harlow: Longman

Tomlin, C. D. and Berry, J. K. (1979) 'A mathematical structure for cartographic modelling in environmental analysis' *Proceedings of the American Congress on Surveying and Mapping*, Falls Church VA: American Congress on Surveying and Mapping

Tomlinson, R. F. (1987) 'Current and potential uses of geographical information systems: the North American experience' *International Journal of GIS* 1, 3: 203–218

Tomlinson, R. F., Calkins, H. W. and Marble, D. F. (1976) *Computer Handling of Geographical Data* Paris: UNESCO Press

Trotter, C. M. (1991) 'Remotely-sensed data as an information source for geographical information systems in natural resource management: a review' *International Journal of GIS* 5, 2: 225–239

Tsai, V. J. D. (1993) 'Delaunay triangulations in TIN creation: an overview and linear-time algorithm' *International Journal of GIS* 7, 6: 501–524

Tufte, E. R. (1990) *Envisioning Information* Cheshire CT: Graphics Press

Unwin, D. (1981) *Introductory Spatial Analysis* London: Methuen

—— (1989) *Curriculum for Teaching Geographical Information Systems* Report to the Education Trust Fund of AUTOCARTO, Royal Institute of Chartered Surveyors, Leicester: University of Leicester

Unwin, T. (1992) *The Place of Geography* Harlow: Longman

van der Knaap, W. G. M. (1992) 'The vector to raster conversion: (mis)use in geographical information systems' *International Journal of GIS* 6, 2: 159–171

Veregin, H. (1989) 'Error modelling for the map overlay operation' in M. F. Goodchild, M. and S. Gopal (eds) *Accuracy of Spatial Databases* London: Taylor and Francis

Vicars, D. (1986) 'Mapdigit – a digitizer interface to GIMMS' *GIMMS Newsletter* 3: 8–9

Visvalingham, M. (1989) 'Cartography, GIS and maps in perspective' *The Cartographic Journal* 26: 25–32

—— (1994) 'Visualisation in GIS, cartography and ViSC' in H. M. Hearnshaw and D. J. Unwin (eds) *Visualization in Geographical Information Systems* Chichester: Wiley

Walker, W., Palimaka, J. and Halustchak, O. (1986) 'Designing a commercial GIS – a spatial relational database approach' *Proceedings of GIS Workshop, Atlanta, GA* Falls Church VA: American Society for Photogrammetry and Remote Sensing

Wang, B. (1986) 'A method to compress raster map data files for storage' in M. Blakemore (ed.) *Proceedings Auto Carto London* vol. 1: 272–281

Webber, R. J. (1980) 'A response to the critique of the OPCS/PRAG national classifications' *Town Planning Review* 51: 440–450

Webster, C. (1988) 'Disaggregated GIS architecture. Lessons from recent developments in multi-site database management systems' *International Journal of GIS* 2, 1: 67–80

——— (1990) *Approaches to Interfacing GIS and Expert System Technologies* Technical Reports in Geo-Information Systems, Computing and Cartography 22, Cardiff: Wales and South West Regional Research Laboratory

Wehde, M. (1982) 'Grid cell size in relation to errors in maps and inventories produced by computerized map processing' *Photogrammetric Engineering and Remote Sensing* 48, 8: 1289–1298

Weibel, R. and Heller, M. (1991) 'Digital terrain modelling' in D. J. Maguire, M. F. Goodchild and D. W. Rhind (eds) *Geographical Information Systems: Principles and Applications* Harlow: Longman

Wells, D. and 11 others (1986) *Guide to GPS Positioning* Ottawa: Canadian GPS Associates

Whitelaw, J. (1986) 'Maps screen Welsh mains' *New Civil Engineer* 3 July: 23

Wiggins, L. L. (1986) 'Three low-cost mapping packages for microcomputers' *APA Journal* Autumn: 480–488

Wilson, P. R. and Elliott, D. J. (1987) 'An evaluation of the Postcode Address File as a sampling frame and its use within OPCS' *Journal of the Royal Statistical Society A* 150: 230–240

Wood, D. (1993) *The Power of Maps* London: Routledge

Wood, M. (1994) 'The traditional map as a visualization technique' in H. M. Hearnshaw and D. J. Unwin (eds) *Visualization in Geographical Information Systems* Chichester: Wiley

Wood, M. and Brodlie, K. (1994) 'ViSC and GIS: some fundamental considerations' in H. M. Hearnshaw and D. J. Unwin (eds) *Visualization in Geographical Information Systems* Chichester: Wiley

Worboys, M. (1994) 'Object-oriented approaches to geo-referenced information' *International Journal of Geographical Information Systems* 8, 4: 385–399

Wrigley, N. (1990) 'ESRC and the 1991 Census' *Environment and Planning A* 22: 573–582

——— (1991) 'Market-based systems of health-care provision, the NHS bill, and geographical information systems' *Environment and Planning A* 23: 5–8

Wyatt, B. K., Briggs, D. and Mounsey, H. M. (1988) 'An information system on the state of the environment in the European Community' in H. M. Mounsey and R. F. Tomlinson (eds) *Building Databases for Global Science* London: Taylor and Francis

Yoeli, P. (1982) 'Cartographic drawing with computers' *Computer Applications 8* University of Nottingham.

——— (1983) 'Digital terrain models and their cartographic and cartometric utilisation' *The Cartographic Journal*, 20, 1: 17–22

——— (1986) 'Computer executed production of a regular grid of height points from digital contours' *The American Cartographer* 13, 3: 219–229

Young, J. A. T. (1986) *A UK Geographic Information System for Environmental Monitoring, Resource Planning and Management Capable of Integrating and Using Satellite Remotely Sensed Data* Monograph 1, Nottingham: The Remote Sensing Society

Zobrist, A. L. (1979) 'Data structures and algorithms for raster data processing' *Proceedings Auto Carto 4* vol. 1: 127–137

INDEX

Page numbers in *italics* refer to figures.